Genomic Uracil

Evolution, Biology, Immunology and Disease

Genomic Uracil

Evolution, Biology, Immunology and Disease

Editors

Geir Slupphaug
Hans Einar Krokan

Norwegian University of Science and Technology, Norway

Foreword by

Nobel Laureate Tomas Lindahl

W World Scientific

W JERSEY · LONDON · SINGAPORE · BEIJING · SHANGHAI · HONG KONG · TAIPEI · CHENNAI · TOKYO

Published by

World Scientific Publishing Co. Pte. Ltd.

5 Toh Tuck Link, Singapore 596224

USA office: 27 Warren Street, Suite 401-402, Hackensack, NJ 07601

UK office: 57 Shelton Street, Covent Garden, London WC2H 9HE

Library of Congress Cataloging-in-Publication Data

Names: Slupphaug, Geir, editor.

Title: Genomic uracil : evolution, biology, immunology and disease / editors,
 Geir Slupphaug, Norwegian University of Science and Technology, Norway,
 Hans Einar Krokan, Norwegian University of Science and Technology, Norway ;
 foreword by Nobel Laureate Tomas Lindahl.

Description: New Jersey : World Scientific, [2018] | Includes bibliographical references and index.

Identifiers: LCCN 2018013174| ISBN 9789813233492 (hardcover : alk. paper) |
 ISBN 9813233494 (hardcover : alk. paper)

Subjects: LCSH: Uracil. | DNA. | Genetics.

Classification: LCC QD401 .G32 2018 | DDC 572.8/8--dc23

LC record available at https://lccn.loc.gov/2018013174

British Library Cataloguing-in-Publication Data

A catalogue record for this book is available from the British Library.

For any available supplementary material, please visit
http://www.worldscientific.com/worldscibooks/10.1142/10803#t=suppl

List of Contributors

Geir Slupphaug, Professor, PhD
Department of Clinical and Molecular Medicine
NTNU-Norwegian University of Science and Technology
NO-7491 Trondheim, Norway; and
PROMEC Core Facility for Proteomics and Metabolomics at
NTNU and the Central Norway Regional Health Authority

Hans E. Krokan, Professor, MD&PhD
Department of Clinical and Molecular Medicine
NTNU-Norwegian University of Science and Technology
NO-7491 Trondheim, Norway

Bodil Kavli, Researcher and Adjunct Professor, PhD
Researcher of The Norwegian Cancer Society
and Adjunct Professor at Department of Clinical and
Molecular Medicine,
NTNU-Norwegian University of Science and Technology
NO-7491 Trondheim, Norway

Finn Drabløs, Professor, PhD
Department of Clinical and Molecular Medicine,
Faculty of Medicine and Health Sciences
NTNU-Norwegian University of Science and Technology
NO-7491 Trondheim, Norway

Pål Sætrom, Professor, DPhil
Department of Computer Science, and
Department of Clinical and Molecular Medicine
NTNU-Norwegian University of Science and Technology
NO-7491 Trondheim, Norway

Cathrine Broberg Vågbø, Senior Scientist, PhD
Department of Clinical and Molecular Medicine
NTNU-Norwegian University of Science and Technology
NO-7491 Trondheim, Norway, and
PROMEC Core Facility for Proteomics and Metabolomics at
NTNU and the Central Norway Regional Health Authority

Antonio Sarno, Postdoctoral Fellow, PhD
Department of Clinical and Molecular Medicine
NTNU-Norwegian University of Science and Technology
NO-7491 Trondheim, Norway, and
The Liaison Committee for Education, Research and Innovation in
Central Norway, Trondheim, Norway

**Henrik Sahlin Pettersen, Associate Professor, and Consultant
Pathologist, MD&PhD**
Department of Clinical and Molecular Medicine
NTNU-Norwegian University of Science and Technology
NO-7491 Trondheim, Norway, and
Department of Pathology, St. Olavs Hospital, Laboratory Centre
NO-7030 Trondheim, Norway

Acknowledgments

The authors have been supported by NTNU-Norwegian University of Science and Technology; The Research Council of Norway; The Norwegian Cancer Society; Svanhild and Arne Must's Fund for Medical Research; The Liaison Committee between the Central Norway Regional Health Authority (RHA) and NTNU; The Cancer Fund at St. Olav's Hospital and Faculty of Medicine and Health Sciences, NTNU.

Contents

Foreword

The replacement of uracil with its methylated derivative, thymine, in relevant polynucleotides was a major step in the generation of a robust carrier of genetic information. However, substantial misincorporation of dUMP residues instead of TMP during DNA replication, as well as mutagenic hydrolytic deamination of cytosine to uracil in DNA, remains as threats to genetic stability and requires continuous correction by DNA editing and DNA repair by special mechanisms. In recent years, it has become clear that such DNA lability is not only a problem of genomic stability but can be of physiological advantage in targeted processing of immunoglobulin genes to provide expanded antibody diversity, and epigenetic oxidative processing of 5-methylcytosine residues during organ differentiation. This makes the study of uracil in DNA of topical interest.

DNA repair research has been a strong field in Norway for decades, thanks to the eminent research groups of Hans Krokan in Trondheim and the late Erling Seeberg in Oslo. An early key contribution from the Krokan group was the isolation from human tissues of large amounts of the low abundance repair enzyme that specifically excises uracil from DNA, the DNA glycosylase UNG. This heroic effort provided the key to the cloning and sequencing of the relevant gene, and subsequent studies of its physiological expression. Hans Krokan, Geir Slupphaug and their key collaborators in Trondheim have continued to make highly relevant contributions to the investigation of the roles of uracil residues in DNA and here provide a broad overview, including the presumed intriguing role of uracil in polynucleotides during the origin of life.

An important aspect of understanding the roles of uracil in DNA concerns the mode of action of the widely used anticancer drug 5-fluorouracil. In a classical achievement of translational research, Charles Heidelberger postulated that a compound such as 5-fluorouracil might interfere with the synthesis of DNA pyrimidine precursors and therefore act as an anticancer agent. When the compound was synthesised and tested, it was found to have profound activity as a useful drug, and it remains a mainstay of drug treatment of several forms of cancer. From this success, it was deduced that the proposed mode of action of 5-fluorouracil had been established, although subsequent studies have only provided controversial support for the model. Recently, Slupphaug and Krokan have obtained new data and propose that the main effect of the drug is due to an entirely different mechanism involving incorporation of 5-fluorouracil residues into DNA to generate toxic base pairs with guanine in the complementary DNA strand. This new approach may improve our understanding of the role of 5-fluorouracil in therapy. It is only one of several recent advances in this area. The present timely volume should be very helpful to understanding the broad implications, positive as well as negative, of the transient occurrence of uracil residues in DNA.

Tomas Lindahl
Nobel Laureate in Chemistry 2015
The Francis Crick Institute, UK

Chapter 1

Introduction — DNA Repair is Integrated with Many Cellular Processes

Hans E. Krokan*, Bodil Kavli and Geir Slupphaug

Long-term survival of a species requires cellular mechanisms that efficiently safeguard DNA, the chemical material of the genome. All known living organisms have several mechanisms that repair different types of DNA damage. This is essential to prevent cytotoxic and mutagenic effects of damage to DNA. In fact, each cell in the body is inflicted by tens of thousands of DNA lesions per day from spontaneous chemical decay and normal cellular metabolites alone.[1] Environmental chemicals, ultraviolet and ionizing radiation add to the burden of genome damage. Without cellular mechanisms that maintain DNA, we would probably die within a few days from multiple organ failure due to inactivation of numerous genes in each cell. Fortunately, a number of mechanisms contribute to maintaining DNA, including high fidelity DNA replication, DNA repair, cell cycle regulation, as well as removal of cells with serious genetic defects by *apoptosis* (programmed cell death) (Fig. 1.1). However, this is just the tip of the iceberg, as outlined below.

* hans.krokan@ntnu.no

1

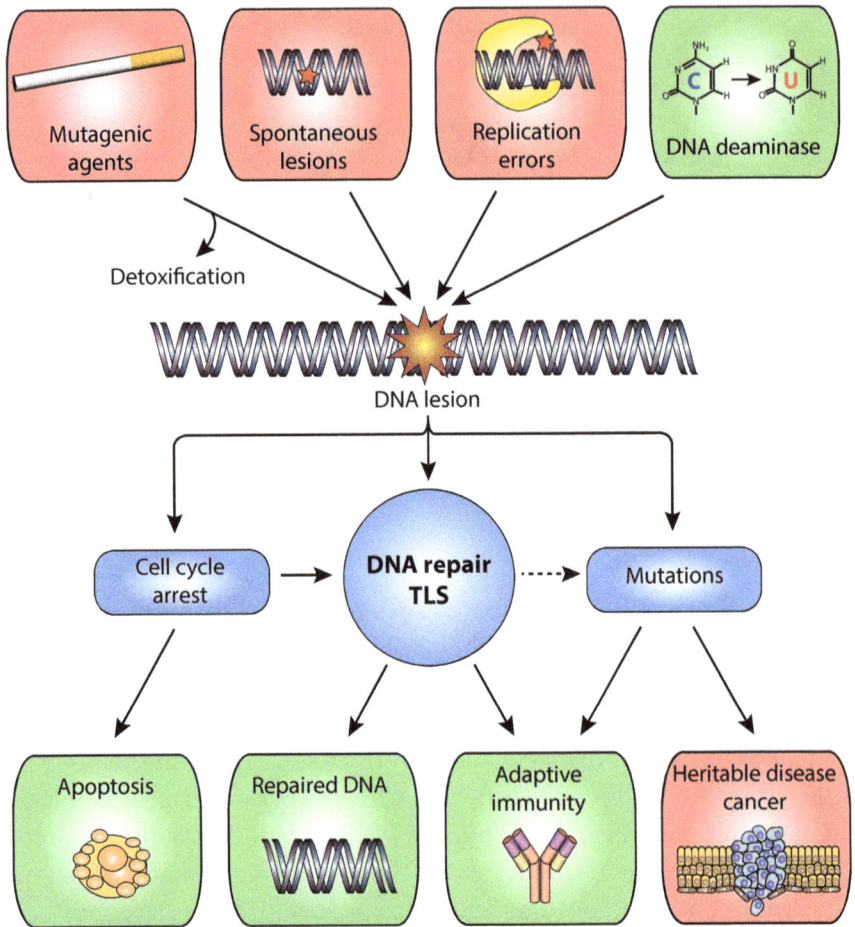

Fig. 1.1. Multiple cellular pathways are coordinated to deal with the plethora of DNA lesions occurring from endogenous and exogenous sources. TLS: translesion synthesis.

1.1. The Genome is Maintained by Integrated Mechanisms

Mechanisms that safeguard DNA do not operate in solitude. Rather, they are highly integrated processes. As one example, DNA damage may arrest the cell cycle progress through signalling via DNA damage checkpoints, giving the cell sufficient time to correct the damage.

This is conceptually rather straightforward. Other mechanisms contribute by detoxifying DNA damaging agents, supplying a balanced pool of DNA precursors, controlling the amount of DNA per cell and ensuring correct segregation of chromosomes during mitosis. All of these processes require regulated transcription, translation, posttranslational modification (PTM) and macromolecule degradation. Frequently, specific PTMs are required to regulate functional interactions of components in repair complexes.

However, recent endeavors in molecular biology and systems biology have also demonstrated that processes previously considered as different entities, are in fact linked by extensive functional networks. For example, several proteins that contribute to regulate inflammatory responses also contribute to regulating DNA repair. A related example is the extensive use of common factors in adaptive immunity and DNA repair. Obviously, we are still very far from understanding the functional complexity of living organisms. In addition, different life forms interact and depend on each other. Life forms are shaped by the "dead" matter such as the chemical composition of Earth that supplies the building material of all life forms and the Sun that supplies the energy for life through short wave length electromagnetic radiation. Chronobiology is the science of biological adaption to cycles of presence (light) and absence (dark) of electromagnetic radiation. Not only does it regulate the periodic requirement for sleep, but light-dependent periodic changes in gene expression also influence the effect of cytostatic drugs, to mention one example. Ultraviolet light has beneficial effects by stimulating biosynthesis of vitamin D_3 in the skin, but it also has detrimental effects as it causes DNA damage of different types, such as thymine dimers that require DNA repair to prevent toxicity and cancer. This is an example of the concept of the principle of duality of various substances and influences, wherein too little and too much may be harmful, and that a "golden mean" was necessary. It was developed by Aristotle (384-322 BC) in his work on ethics (Nicomachean Ethics; named after his son Nicomachus). It applies to many aspects of biology and medicine. The great Italian renaissance artist, philosopher, scientist and inventor,

Leonardo da Vinci (1452–1519), developed this thinking further. He is the source of the quote, "To develop a complete mind: Study the art of science; study the science of art. Learn how to see. Realize that everything connects to everything else." Leonardo da Vinci is perhaps the best example of a universal genius and the cited quote is universally applicable, including the understanding of the role of genome maintenance in health and disease.

1.2. What is DNA Repair and Why is It So Important?

DNA repair corrects DNA lesions or misplaced bases in DNA, thereby restoring the normal DNA sequence.[2] The publication of the structure of DNA in 1953 by James Watson and Francis Crick[3] suggested a mechanism of DNA replication involving copying of the strand based on the principles of complementarity. However, DNA repair was not considered then. In 1974, Francis Crick stated the following in a paper celebrating the first 21 years of the double helix: "We totally missed the possible role of enzymes in repair although, due to Claude Rupert's early very elegant work on photoreactivation, I later came to realize that DNA is so precious that probably many distinct repair mechanisms would exist. Nowadays one could hardly discuss mutation without considering repair at the same time."[4] At this time James Cleaver had already demonstrated that deficient DNA repair causes *Xeroderma pigmentosum*, a hallmark of which is multiple skin cancers from exposure to ultraviolet light.[5] Today, it is generally accepted that the fundamental cause of cancer is mutations caused by DNA damage that has not been corrected. While preventing mutations and cancer is certainly an important long-term function of DNA repair, alleviating cytotoxicity from stalled replication and transcription at sites of DNA damage is also very important. It is in fact essential for the survival of cells, even in a very short-term perspective. It is also probably the evolutionary oldest function of DNA repair, since it predates multicellular organisms in evolution. Although cancer is caused by failure to correct

DNA damage, *enhanced* repair capacity may sometimes give cancer cells growth advantages and resistance to cytostatic drugs. As a consequence of this, some DNA repair proteins are also interesting cancer drug targets.

1.3. Mechanisms of DNA Repair

DNA repair takes place by several different mechanisms. The first definite demonstration of DNA repair was light-dependent enzymatic repair of DNA damage caused by ultraviolet light, pioneered by Claud Stanley Rupert in 1960. Subsequently, he characterized the responsible enzyme, photolyase.[6,7] The substrate of this enzyme was shown to be pyrimidine dimers, which are covalently linked adjacent pyrimidines into the same DNA strand. Photolyase converts dimers to two normal pyrimidines in a one-step, light-dependent reaction. Certain alkylation lesions in DNA are also repaired in one step by single proteins, including O6-guanine alkyltransferase and at least two oxidative dealkylases that remove certain alkylation lesions in purines and pyrimidines in DNA. Such one-step repair mechanisms correct DNA bases directly, without sequence information from the complementary strand. Photolyase is present in prokaryotes and eukaryotes, including marsupial mammals, but not placental mammals. Alkyltransferases and oxidative dealkylases are evolutionary conserved enzymes present in prokaryotes and eukaryotes, including man. However, one-step direct repair proteins have a limited repertoire relative to the vast number of DNA lesions known. These are mostly processed by excision repair mechanisms. They have in common that the damaged nucleotide (and frequently surrounding nucleotides) is removed rather than repaired, thus creating a single-stranded gap in DNA. The gap is then filled and sealed by DNA synthesis and ligation, using sequence information from the complementary strand to insert the correct nucleotide. DNA excision repair always takes place by multi-step mechanisms, the classical ones being base excision repair (BER), single-strand break repair (SSBR), nucleotide excision repair (NER) and mismatch repair (MMR). In addition, double-strand

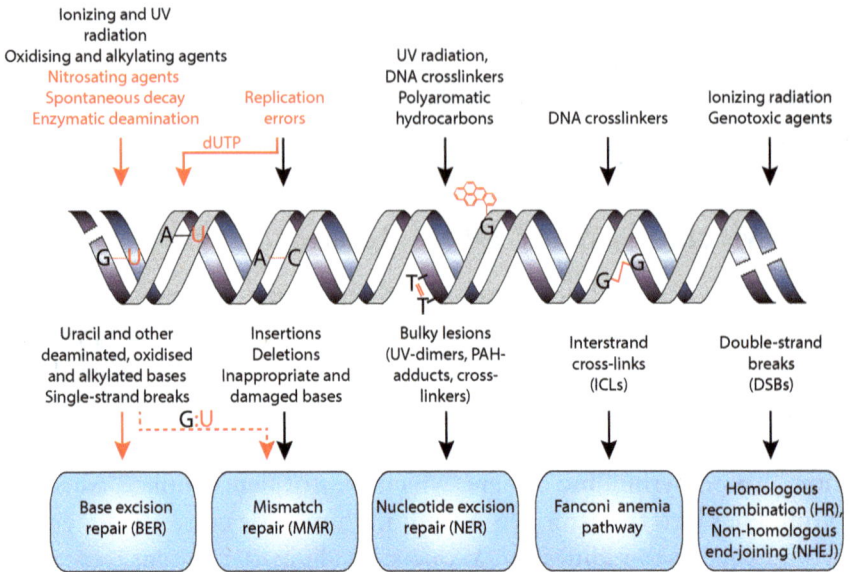

Fig. 1.2. The different DNA excision repair pathways in mammalian cells. Pathways related to genomic uracil are highlighted by red arrows. Repair of U in G:U mismatches may also be repaired by MMR, possibly as a last resort. Furthermore, some replication errors, e.g. misincorporated A opposite template G and T opposite G can be repaired by BER. Several examples of intra-pathway and inter-pathway complementation are known. Homologous recombination is included among the DSBR pathways, although DNA damage excision does not occur during this type of repair.

break repair (DSBR) also requires multi-step pathways. For each of these repair pathways, there are sub-pathways. An overview of the different excision repair pathways and their molecular functions is presented in Fig. 1.2.

1.4. Uracil in DNA — Old Story with Renewed Interest

Given the plethora of lesions in DNA, is it justified to single out one such lesion for an extensive review in the form of a book? We believe that the following chapters will answer this question more completely. The short answer is that uracil in DNA is both a general mutagenic lesion and a cell-specific intermediate in the immunoglobulin genes of B-cells during antibody maturation. Generally, it is found as U:G mismatches and U:A pairs, depending on its origin. U:G mismatches arise from spontaneous or enzymatic deamination of cytosine in DNA by members of the AID/APOBEC family. Importantly, such mismatches seem to give rise to a substantial fraction of mutations observed in many common forms of human cancer. U:A pairs arise from incorporation of dUMP during replication and may be weakly mutagenic through abasic site intermediates in repair. Genomic uracil is mostly processed by error-free mechanisms, but at least in B-lymphocytes also by mutagenic mechanisms. It may even have epigenetic functions related to cytosine demethylation. It has now become apparent that cytosine, the most unstable of DNA bases, is used for several functions linked to cytosine 5-methylation or deamination. However, the usefulness of this instability comes with a risk of untargeted deamination and mutagenesis.

1.5. Why Do Cells Have Thymine Instead of Uracil in DNA?

All cells normally have thymine rather than uracil in their DNA (Fig. 1.3), but as mentioned, low levels of uracil occur in genomes due to deamination of cytosine and incorporation of dUMP from dUTP, which is normally present in very low concentrations in the cell. Bacterial and mammalian DNA polymerases use dUTP and dTTP with essentially equal efficiency.[8,9]

Furthermore, the structure of DNA containing uracil in U:A pairs is equal to that containing T:A pairs.[10] Thus, there is no obvious biochemical reason why uracil could not work as carrier of genetic information, except that the cell could not easily discriminate between "normal" uracil and uracil resulting from deamination of cytosine. The cell has solved the stability problem by using thymine, which chemically is 5-methyluracil, instead of uracil. This allows the more labile cytosine to be used for different purposes. Uracil in DNA is rapidly removed by base excision repair (BER), initiated by

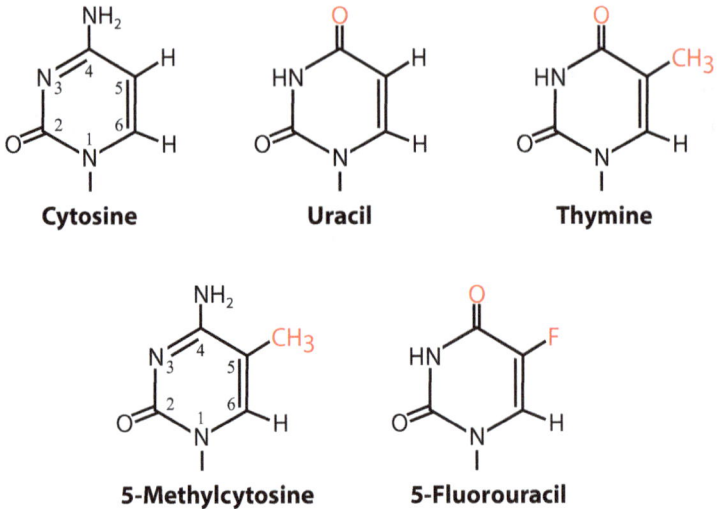

Fig. 1.3.   Chemical structures of cytosine, uracil, thymine and 5-metnylcytosine, which are all found to be constituents of normal DNA (uracil in activated B-cells). Shown is also 5-fluorouracil, which is incorporated to significant levels in DNA after chemotherapy with 5-flourouracil or 5-fluorodeoxyuridine. The molecular substitutions (red) compared to cytosine are located exclusively at the Watson-Crick base pairing positions at C4 and C5.

a uracil-DNA glycosylase. Bacteria carrying mutations that strongly increase dUMP-incorporation in the genome are growth arrested.[11] Thus, cells have adapted to a life with thymine in the genome and do not tolerate large amounts of U:A pairs. However, the small uracil-containing DNA bacteriophage PBS2 can survive in *Bacillus subtilis* because it encodes a peptide inhibitor (Ugi) of uracil-DNA glycosylase.[12]

1.6. Sources and Processing of Genomic Uracil — An Overview

Spontaneous deamination of DNA-cytosine to uracil[13] and incorporation of dUMP instead of dTMP[9,14,15] were reported approximately four decades ago. The Ung-type uracil-DNA glycosylase was discovered during a search for an enzyme that would act on U:G mismatches; the highly mutagenic lesion resulting from spontaneous deamination of cytosine in DNA. This remarkable discovery also marks the discovery of base excision repair (BER) as a distinct mechanism.[16] Subsequently, a number of other DNA glycosylases were discovered, demonstrating the versatility of BER as a DNA repair mechanism.

More recently, it has been established that genomic uracil may also result from targeted and untargeted enzymatic deamination of cytosine in DNA by activation-induced cytosine deaminase (AID) or other members of the APOBEC-family. These relatively uncommon events are in most cells efficiently corrected by BER initiated by a uracil-DNA glycosylase (Fig. 1.4). In mammalian cells, four different uracil-DNA glycosylases are known, including nuclear UNG2 and mitochondrial UNG1 (both encoded by the *UNG* gene), SMUG1, TDG and

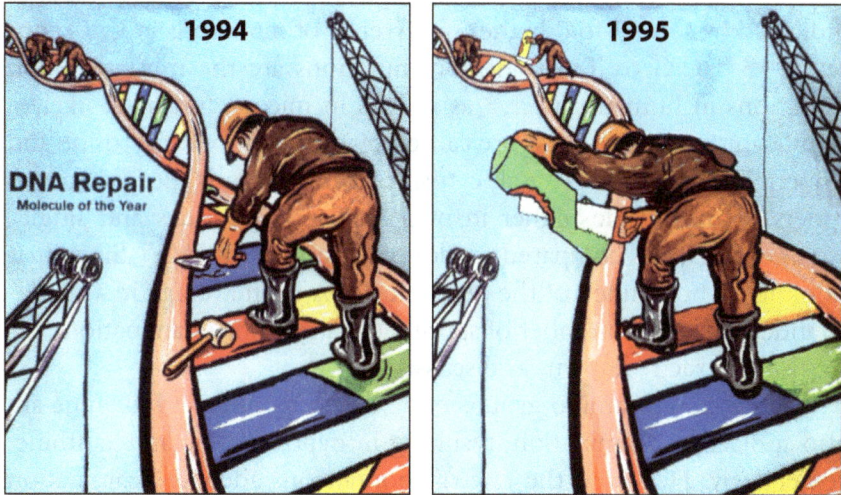

Fig. 1.4. Artistic view of DNA repair. The left panel is an illustration by R.J. Kaufman at the front page of *Science* December 23, 1994. The right panel is an adapted illustration from Tidsskr Nor Legefor, issue 13, 1998, modified by the authors (with permission from the artist) after the discovery that uracil base excision repair is initiated by flipping the uracil out of the double helix prior to cleavage of the N-glycosidic bond by UNG.[17-19]

MBD4. These enzymes have different biochemical properties and expression patterns and have only in part overlapping functions.[20–25]

Incorporated dUMP may be the most abundant source of genomic uracil. The resulting U:A pairs are not miscoding, but may cause mutations by translesion synthesis at abasic sites generated by uracil-DNA glycosylase[26] or from errors in gap-filling in the BER process.[27,28] Cytosine deamination, resulting in U:G mismatches,[1,13,16] is more intriguing since copying of U during replication will invariably result in the generation of a U:A pair, and eventually a T:A pair, in the place of the original C:G pair. Repair of U:G mismatches is therefore essential to avoid mutations. Furthermore, 5-methylcytosine in CpG-sequences, a post-replicative cytosine-modification that regulates gene expression, is also subject to deamination, resulting in mutagenic T:G mismatches.[29,30] The rate of spontaneous deamination of 5-methylcytosine in double-stranded DNA is approximately twice as high as that of cytosine.[31] In addition, deaminated 5-methylcytosine residues (T:G pairs) are apparently less efficiently repaired, relative to U:G pairs, as indicated by several-fold higher occurrence of mutations in CpG contexts.[30,32–34] C:G to T:A transition mutations are the most common mutations in human cancer,[35] as well as in mice,[36] and most likely a significant fraction of these are caused by deamination of cytosine and 5-methylcytosine. In general, cytosine deamination is more than two orders of magnitude higher in nucleosides, nucleotides and single-stranded DNA as compared to double-stranded DNA.[1,37] Since it is not known how much of the genomic DNA is on average in a single-stranded form, the number of spontaneous cytosine deaminations per day is not exactly known, as discussed.[20]

The exocyclic amino groups of genomic adenine and guanine are also subject to deamination, resulting in hypoxanthine and xanthine, respectively. However, the rate of spontaneous adenine deamination in DNA is some two orders of magnitude lower than that of cytosine.[1,37] The rate of spontaneous deamination of guanine in DNA to xanthine is apparently not known, but assumed to be low. However, nitrosative deamination in DNA of adenine and guanine, as well as cytosine, may be significant and medically important, particularly during inflammation. Here nitric oxide (NO·) and nitrosative derivatives such as nitrous anhydride and peroxynitrite are formed in substantial amounts.[38] Deamination by peroxynitrite is apparently restricted to guanine.[39]

Genomic cytosine is rather unique among DNA bases due to its chemical instability and many enzymatic modifications of great significance for understanding DNA repair, gene regulation, innate immunity, adaptive immunity and cancer development, as recently reviewed.[23,40–42] This book focuses on generation and processing of genomic uracil in mammalian cells and its biological significance.

References

1. Lindahl, T. Instability and decay of the primary structure of DNA. *Nature* **362**, 709–715 (1993).
2. Friedberg, E. C., Siede, W., Wood. R. D., *et al. DNA Repair and Mutagenesis.* (ASM Press, 2006).
3. Watson, J. D., Crick, F. H. Molecular structure of nucleic acids; a structure for deoxyribose nucleic acid. *Nature* **171**, 737–738 (1953).
4. Crick, F. The double helix: a personal view. *Nature* **248**, 766–769 (1974).
5. Cleaver, J. E. Defective repair replication of DNA in xeroderma pigmentosum. *Nature* **218**, 652–656 (1968).
6. Rupert, C. S. Photoenzymatic repair of ultraviolet damage in DNA. I. Kinetics of the reaction. *J Gen Physiol* **45**, 703–724 (1962).
7. Rupert, C. S. Photoenzymatic repair of ultraviolet damage in DNA. II. Formation of an enzyme-substrate complex. *J Gen Physiol* **45**, 725–741 (1962).
8. Bessman, M. J., Lehman, I. R., Adler, J., *et al.* Enzymatic Synthesis of Deoxyribonucleic Acid. Iii. The incorporation of pyrimidine and purine analogues into deoxyribonucleic acid. *Proc Natl Acad Sci U S A* **44**, 633–640 (1958).
9. Wist, E., Unhjem, O., Krokan, H. Accumulation of small fragments of DNA in isolated HeLa cell nuclei due to transient incorporation of dUMP. *Biochim Biophys Acta* **520**, 253–270 (1978).
10. Langridge, R., Marmur, J. X-Ray diffraction study of a DNA which contains uracil. *Science* **143**, 1450–1451 (1964).
11. el-Hajj, H. H., Wang, L., Weiss, B. Multiple mutant of Escherichia coli synthesizing virtually thymineless DNA during limited growth. *J Bacteriol* **174**, 4450–4456 (1992).
12. Friedberg, E. C., Ganesan, A. K., Minton, K. N-Glycosidase activity in extracts of Bacillus subtilis and its inhibition after infection with bacteriophage PBS2. *J Virol* **16**, 315–321 (1975).

13. Lindahl, T., Nyberg, B. Heat-induced deamination of cytosine residues in deoxyribonucleic acid. *Biochemistry* **13**, 3405–3410 (1974).

14. Tye, B. K., Nyman, P. O., Lehman, I. R., *et al.* Transient accumulation of Okazaki fragments as a result of uracil incorporation into nascent DNA. *Proc Natl Acad Sci USA* **74**, 154–157 (1977).

15. Grafstrom, R. H., Tseng, B. Y., Goulian, M. The incorporation of uracil into animal cell DNA in vitro. *Cell* **15**, 131–140 (1978).

16. Lindahl, T. An N-glycosidase from Escherichia coli that releases free uracil from DNA containing deaminated cytosine residues. *Proc Natl Acad Sci USA* **71**, 3649–3653 (1974).

17. Mol, C. D., Arvai, A. S., Slupphaug, G., *et al.* Crystal structure and mutational analysis of human uracil-DNA glycosylase: structural basis for specificity and catalysis. *Cell* **80**, 869–878 (1995).

18. Savva, R., McAuley-Hecht, K., Brown, T., Pearl, L. The structural basis of specific base-excision repair by uracil-DNA glycosylase. *Nature* **373**, 487–493 (1995).

19. Slupphaug, G., Mol, C. D., Kavli, B., *et al.* A nucleotide-flipping mechanism from the structure of human uracil-DNA glycosylase bound to DNA. *Nature* **384**, 87–92 (1996).

20. Kavli, B., Otterlei, M., Slupphaug, G., Krokan, H. E. Uracil in DNA--general mutagen, but normal intermediate in acquired immunity. *DNA Repair (Amst)* **6**, 505–516 (2007).

21. Krokan, H. E., Bjoras, M. Base excision repair. *Cold Spring Harb Perspect Biol* **5**, a012583 (2013).

22. Krokan, H. E., Drablos, F., Slupphaug, G. Uracil in DNA--occurrence, consequences and repair. *Oncogene* **21**, 8935–8948 (2002).

23. Krokan, H. E., Saetrom, P., Aas, P. A., *et al.* Error-free versus mutagenic processing of genomic uracil-Relevance to cancer. *DNA Repair (Amst)* **19**, 38–47 (2014).

24. Lindahl, T. My journey to DNA repair. *Genomics Proteomics Bioinformatics* **11**, 2–7 (2013).

25. Olinski, R., Jurgowiak, M., Zaremba, T. Uracil in DNA--its biological significance. *Mutat Res* **705**, 239–245 (2010).

26. Auerbach, P., Bennett, R. A., Bailey, E. A., Krokan, H. E., Demple, B. Mutagenic specificity of endogenously generated abasic sites in Saccharomyces cerevisiae chromosomal DNA. *Proc Natl Acad Sci U S A* **102**, 17711–17716 (2005).

27. Akbari, M., Pena-Diaz, J., Andersen, S., *et al.* Extracts of proliferating and non-proliferating human cells display different base excision pathways and repair fidelity. *DNA Repair (Amst)* **8**, 834–843 (2009).

28. Bennett, S. E., Sung, J. S., Mosbaugh, D. W. Fidelity of uracil-initiated base excision DNA repair in DNA polymerase beta-proficient and -deficient mouse embryonic fibroblast cell extracts. *J Biol Chem* **276**, 42588–42600 (2001).

29. Coulondre, C., Miller, J. H., Farabaugh, P. J., Gilbert, W. Molecular basis of base substitution hotspots in Escherichia coli. *Nature* **274**, 775–780 (1978).

30. Duncan, B. K., Miller, J. H. Mutagenic deamination of cytosine residues in DNA. *Nature* **287**, 560–561 (1980).

31. Shen, J. C., Rideout, W. M., 3rd, Jones, P. A. The rate of hydrolytic deamination of 5-methylcytosine in double-stranded DNA. *Nucleic Acids Res* **22**, 972–976 (1994).

32. Campbell, C. D., Chong, J. X., Malig, M., *et al.* Estimating the human mutation rate using autozygosity in a founder population. *Nat Genet* **44**, 1277–1281 (2012).

33. Meng, H., Cao, Y., Qin, J., *et al.* DNA methylation, its mediators and genome integrity. *Int J Biol Sci* **11**, 604–617 (2015).

34. Rideout, W. M., 3rd, Coetzee, G. A., Olumi, A. F., Jones, P. A. 5-Methylcytosine as an endogenous mutagen in the human LDL receptor and p53 genes. *Science* **249**, 1288–1290 (1990).

35. Kandoth, C., McLellan, M. D., Vandin, F., *et al.* Mutational landscape and significance across 12 major cancer types. *Nature* **502**, 333-339 (2013).

36. Stuart, G. R., Oda, Y., de Boer, J. G., Glickman, B. W. Mutation frequency and specificity with age in liver, bladder and brain of lacI transgenic mice. *Genetics* **154**, 1291–1300 (2000).

37. Shapiro, R. in *Chromosome Damage and Repair, Ed's Erling Seeberg and Kjell Kleppe* Vol. 40, 3–18 (Plenum Press, ISBN 0-306-40886-4, New York London, 1980).

38. Dedon, P. C., Tannenbaum, S. R. Reactive nitrogen species in the chemical biology of inflammation. *Arch Biochem Biophys* **423**, 12–22 (2004).

39. Burney, S., Caulfield, J. L., Niles, J. C., *et al.* The chemistry of DNA damage from nitric oxide and peroxynitrite. *Mutat Res* **424**, 37–49 (1999).

40. Knisbacher, B. A., Gerber, D., Levanon, E. Y. DNA Editing by APOBECs: A Genomic Preserver and Transformer. *Trends Genet* **32**, 16–28 (2016).

41. Harris, R. S., Dudley, J. P. APOBECs and virus restriction. *Virology* **479–480**, 131–145 (2015).

42. Hashimoto, H., Zhang, X., Vertino, P. M., Cheng, X. The Mechanisms of Generation, Recognition, and Erasure of DNA 5-Methylcytosine and Thymine Oxidations. *J Biol Chem* **290**, 20723–20733 (2015).

Chapter 2

The Earth, Life and Genomic Uracil

Hans E. Krokan*, Bodil Kavli, Geir Slupphaug
and Finn Drabløs

Life emerged some 3.8 billion years ago, possibly even earlier. Furthermore, in the scenarios for non-enzymatic synthesis of molecules still present in all life forms, uracil was among those found in substantial yields. Uracil may thus have been one of the original bases in RNA polymers in the hypothetical prebiotic RNA World. It may also have been one of the bases in the first cells, which were thought to contain RNA as the hereditary material. The transition to a DNA World had major advantages; the DNA backbone is much more stable than that of RNA because of the reactivity of 2′-hydroxyl group in ribose. This allowed expansion of DNA-genome sizes. Furthermore, RNA replication is intrinsically error-prone compared to DNA replication.[1] DNA cells appear to have completely outcompeted RNA cells. If RNA cells still exist, they must be confined to hitherto unexplored niches. RNA viruses may be remnants of RNA cells, although their origin remains uncertain. The transition from RNA to DNA cells was made possible by evolution of ribonucleotide reductase (RNR, Chapter 3.1.3). This enzyme converts ribonucleotides to deoxyribonucleotides, the building blocks of DNA. The first DNA cells probably contained uracil in the genome, since RNR

---

* hans.krokan@ntnu.no

could directly produce dATP, dGTP, dCTP and dUTP from the corresponding ribonucleotides, but not dTTP. Furthermore, DNA-viruses carrying uracil instead of thymine in their genomes do exist, possibly representing relics of a uracil-DNA world.

The transition to a thymine-DNA genome had a major advantage; genomic uracil arising from misincorporation of dUTP or deamination of cytosine, the least stable base in DNA, would stand out as an anomalous base requiring removal and correction. The replacement of genomic uracil by thymine was made possible by evolution of a dUTPase, a thymidylate synthase, and finally a uracil-DNA glycosylase to remove traces of genomic uracil, which could arise from cytosine deamination and incorporation of dUTP. Thus, genomic uracil, nucleotide metabolizing enzymes and DNA correction systems must have been intimately linked to early steps in the evolution of cellular life, possibly emerging in the temporal order mentioned. Taking a broad evolutionary approach, we will explore these topics further below.

2.1. Life on Earth has Existed for More Than 80% of its Age

The Earth is ~4.6 billion years old. Our planetary system is thought to have arisen from condensation of solar nebular (gaseous) material surrounding what was to become our sun. The Moon-forming impact, thought to result from a collision of the Earth with an astronomical body the size of Mars ~4.5 billion years ago, caused extreme heating that precluded evolution or continuation of life. This phase was followed by gradual cooling and establishment of a greenhouse atmosphere through complex chemical processes that eventually stabilized the temperature in atmosphere and surface.[2] It is thought that life on the planet started during the early Archean era ~3.8 billion years ago with an ancestral primitive prokaryotic organism. Most likely several different cell-like organisms emerged and disappeared before arrival of LUCA, the Last Universal Common Ancestor. The age of the oldest eukaryotes is not known, but evidence from fossils and molecular clocks suggests one to two billion years.[3,4] The "last eukaryote common ancestor" (LECA) had most likely acquired mitochondria by endosymbiosis of an α-proteobacterial genome prior to diversification to extant phyla. The eukaryote host in this endosymbiotic process was most likely related to Archaea.[5] Multicellular eukaryote life apparently evolved independently in different clades some 600 million years ago.[6,7] This was followed by relatively rapid evolution of larger and more complex animals. The oldest vertebrate identified, a primitive jawless fish, thus evolved more than 500 million years ago.[8,9] Other vertebrates, including jawed cartilaginous fish, bony fish and reptiles evolved some 100 million years later. The first mammals did not evolve until 200 million years ago, with egg-laying (Prototheria) and marsupial mammals (Marsupialia) preceding placental mammals.[10,11] Modern man and our closest living relative, the chimpanzee, diverged approximately 6–7 million years ago, but have very similar genomes. Only 35 million single nucleotide changes overall (~1%) were found and the pattern of evolution of protein-coding genes are strongly correlated in the two species.[12] However, sequencing and more detailed analyses of numerous human and non-human genomes have revealed many differences in developmental genes and regulatory regions.[13] The first *Homo* genus, *Homo habilis*, appeared 2.8 million years ago. The ancestor of the present-day humans and our closest relatives, the Neanderthals and Denisovans, split 500,000 to 700,000 years ago. The modern human, *Homo sapiens sapiens*, appeared as late as 200,000 years ago, and has thus existed less than 0.005% of the history of Earth. Although all other homo species are extinct, modern man carries a small fraction (~1.5–2.5%) of Neanderthal DNA in the genome. An early modern human that died 37,000–42,000 years ago carried as much as 6–9% Neanderthal DNA, indicating that this individual had a Neanderthal ancestor only four to six generations back.[14] The presence of Neanderthal DNA in modern humans, notably except in sub-Saharan Africa, indicates that interbreeding did occur. This apparently took place approximately 40,000 years ago, but may have been of limited extent.[14,15] Recently, it was shown that ancestors of Neanderthals from the Altai Mountains (where Denisovans were first identified) and modern man interbred many thousand years earlier than thought previously.[16] What caused extinction of Neanderthals is unknown; it could be disease, climate changes, violent clashes with modern man or perhaps they were outnumbered and absorbed into modern man through interbreeding.

Sequencing of ancient DNA has become a central element in studies on our evolution as a species. One major problem in such analyses is contamination of homo DNA by microorganisms and extensive decay of the ancient DNA, including deamination of cytosine to uracil. However, uracil-containing DNA oligonucleotides have also become a tool in molecular biology. Chapter 9 discusses DNA-uracil as a molecular biology tool, as well as a problem in analyses of ancient DNA. This chapter also briefly discusses why claimed analysis of DNA much older than ~100,000 years most likely is fiction.

2.2. The RNA World — Prebiotic Self-replication of Macromolecules — Uracil was There!

Cells are live organisms with metabolism. During evolution, cells must have been preceded by a simpler form of self-replicating molecules. According to the "RNA World" hypothesis, these were RNA molecules (Fig. 2.1) generated from smaller chemicals in the "primordial soup" as reviewed.[17–19] During and after the cooling period following the Moon-forming impact, the temperatures and chemistry of the atmosphere and surface of the Earth were very different than today. The RNA World hypothesis is attractive because RNA can carry genetic information *and* function as a catalyst. Naturally, we lack direct evidence for the existence of ribozymes in the prebiotic era. Although most ribozymes are not abundant and the number of different ribozymes in cells today is limited, they may have been more abundant and diversified early in evolution. Furthermore, strongly improved activity of *engineered* variants of an RNA polymerase ribozyme demonstrated their capacity to produce a variety of functional RNAs.[20] Other ribozyme variants could carry out both phosphoryl and nucleotidyl transfer as well as nucleobase modifications. These findings demonstrate the potential and versatility of ribozymes.[21]

(A) **RNA constituents**

(B) **Hairpin ribozyme**

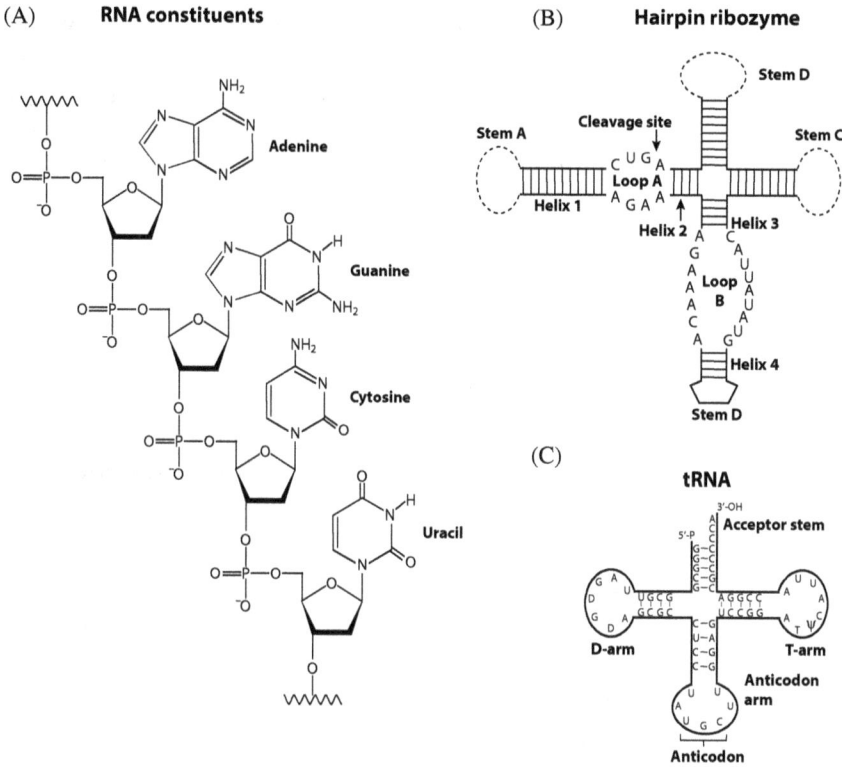

(C) **tRNA**

Fig. 2.1. RNA molecules are diverse in chemical structure and composition. (A) A textbook type representation of the chemical structure of canonical bases and backbone ribose-phosphate. (B) 2D structure of a hairpin ribozyme with helix stems and loops. (C) A general 2D cloverleaf structure of transfer RNA (tRNA), which contains many base modifications, e.g. dihydrouridine (D) in the D-arm (or D-loop), pseudouridine (ψ), thymine, as well as methylations in C, A and G (not shown).

In a transition period preceding polymerization by RNA ribozymes, non-enzymatic copying of RNA templates involving short activated oligomers would seem plausible.[22]

In support of the RNA World hypothesis, many chemical compounds that are building blocks of RNA and proteins are present in carbonaceous meteorites. These are thought to resemble the composition of the very young Earth, before life emerged. The most famous of such meteorites is the Murchison meteorite, named after the

Australian town where this large, fragmented meteorite fell down in 1969. Fragments of the Murchison meteorite, and many smaller ones, have been extensively studied. The content of carbon-containing molecules in meteorites sometimes exceeds 10% and includes several amino acids now found in proteins, but with near equal amounts of *d*- and *l*-forms, as would be expected for non-enzymatic synthesis. Other carbon-molecules include uracil and purines, hydroxyacids, ketoacids, dicarboxylic acids, sugar alcohols, aldehydes, ketones and numerous hydrocarbons, as reviewed,[17] but apparently not cytosine. Thus, the four present-day RNA or DNA bases, as well as other required compounds, may not have been readily available in sufficient concentrations to generate self- replicating molecules. However, even if the full complement of RNA/DNA building blocks was not available, this does not really argue against the generation of a simpler form of RNA/DNA-like molecules, later to be replaced by the present-day type of DNA/RNA.

More recently, a formamide-based scenario for prebiotic generation of components known from fundamental biochemistry, including a full complement of nucleobases, has attracted much attention (Fig. 2.2). In the presence of powdered meteorites of different types as catalysts and energy from e.g. proton radiation, formamide is a starting point for generation of all RNA-components (bases, nucleosides and nucleotides), molecules of the tricarboxylic acid cycle, fatty acids, sugars and amino acids.[24-26] Furthermore, during the "last heavy bombardment period" (LHB) 4.1–3.8 billion years ago, numerous large comets and asteroids hit the Earth, causing extreme heat in large areas. In this period, a wide range of small molecules relevant to life may have been generated in substantial amounts. Simulating LHB-conditions using a high-energy laser, the abundant and simple molecule formamide was split into CN and NH radicals that again reacted with formamide to generate RNA nucleobases and nucleosides.[27] In all the described scenarios, uracil was among the compounds generated in good yield. Interestingly, the LHB period coincides with the period during which life emerged.

The many abilities of RNA would seem to add plausibility to its possible function as a prebiotic information and catalytic molecule.

From simple molecules to the RNA world

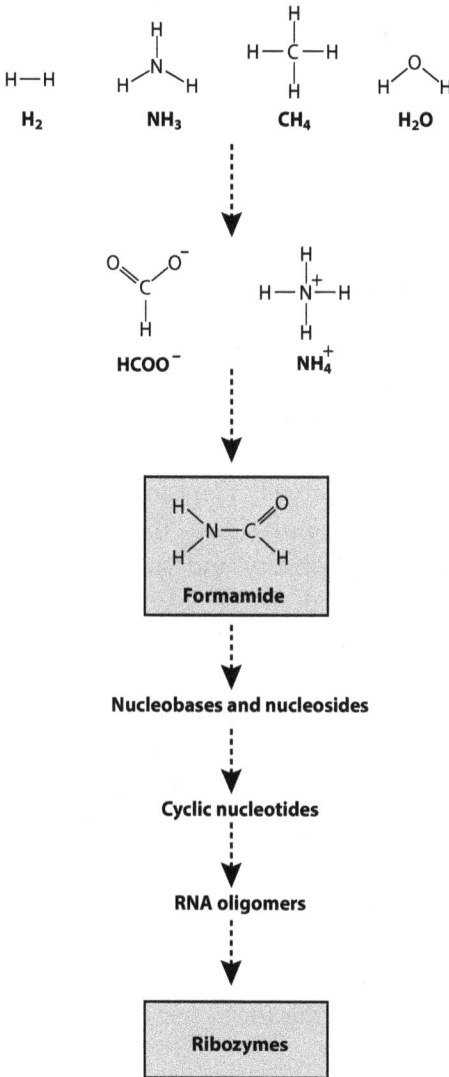

Fig. 2.2. From simple molecules to the RNA World. Prebiotic synthesis of components found in RNA — the formamide scenario. In the presence of small chemicals and input of high energy in the form of heat and radiation, all RNA components have been shown to be generated from formamide. It should be underlined that this is one of several scenarios for generation of chemicals now found in living cells.

These abilities include non-enzymatic template-directed polymerisation, template-directed RNA polymerisation *in vitro* by an RNA polymerase ribozyme, *in vitro* evolution of viral RNA by environmental selection and regulation of gene expression by riboswitches. Accumulation of RNA and other molecules in cell-like lipid vesicles, could concentrate reactants and speed up spontaneous and RNA-catalysed chemical reactions; as reviewed.[19]

2.3. A Less Pure RNA World?

Although the RNA world hypothesis is attractive and plausible, it has been challenged and may require modifications and additions. Amino acids and short peptides may have had a decisive role from the beginning. An early form of RNA may have selectively interacted with amino acids in a process eventually resulting in code-instructed peptide synthesis. Binding of amino acids to the acceptor stem loop may have depended on hydrophobicity and size of the amino acids, as indicated from work on "modern" tRNAs. The anticodon loop may have developed later to take on a role in recognizing peptide structure. Thus, there are two identity elements for each amino acid; one in the acceptor stem and the other in the anticodon loop.[28,29] Selective interactions of amino acids with a postulated ancestral tRNA-like molecule would then be one clue to understanding the origin of the genetic code. In other words, early life evolved from an RNA-peptide World, rather than a "pure" RNA World. Others have also argued that an RNA replicator could not have been the origin of life, instead favouring early metabolism in the form of a set of energy-dependent non-enzymatic catalytic reactions between smaller molecules, with RNA/DNA replicators entering the scene only later.[30]

2.4. Old RNA Viruses Have Limited Similarity to Modern RNA Viruses

RNA viruses carry uracil instead of thymine in their genomes. It would seem plausible that RNA viruses could emerge from self-replicating RNA molecules prior to the emergence of cellular life. However, the origin of viruses is still debated. Since viruses depend on a host for their propagation, one point of view is that they did not come first, but have escaped from their host cell later in evolution. This would apply to both RNA and DNA viruses. Due to their high mutation rates, the genomes of RNA viruses present today are probably very different from the ancient ones, if they existed. The views on the origin of RNA viruses are indeed conflicting. Using a molecular clock approach to study the age of RNA viruses, a method that takes into account mutation rates and sequence divergence of related viruses, the RNA virus families present today have originated very recently, perhaps only 50,000 to 100,000 years ago. However, while this approach may be useful to determine the age and common origin of families present today, it does not really give information on how old RNA viruses are. Other types of evidence, such as cospeciation and virus morphology, indicate a much more ancient origin.[31,32] It is also possible that RNA viruses may have entered the scene more than once; the first time preceding cellular life into which it may have evolved as an RNA-genome cell type, later to be replaced by a DNA-genome cell. Further on in evolution RNA and DNA may have "escaped" from the DNA-genome cell, forming new RNA and DNA viruses. In contrast to the rapidly evolving genome of RNA viruses, their protein structures are much better conserved and generally suggest an ancient origin. Such studies strongly suggest that both RNA and DNA viruses are ancient, with RNA viruses preceding DNA viruses.[33–36]

2.5. Genomic Uracil in RNA Viruses and in Some Bacterial DNA Viruses

Uracil is carrier of genetic information, together with cytosine adenine and guanine in RNA bacteriophages (bacterial RNA viruses), numerous eukaryotic RNA viruses, as well as vertebral retroviruses. Generally, eukaryotic RNA viruses are much more abundant than prokaryotic ones. The retroviruses are usually not classified as RNA viruses, although they carry RNA in their genomes. This is because retroviral replication uses a DNA replication intermediate (proviral DNA) as template for production of progeny RNA. Importantly, human endogenous retroviruses (HERVs) comprise as much as 8% of the human genome and have been important in the evolution of vertebral genomes.[37] Although largely inactivated by mutations during evolution, some have important regulatory functions and a few are even expressed at the protein level, particularly in cancer cells.[38]

No known present-day cellular organisms use uracil as information-carrier in their genome. However, *Bacillus subtilis* bacteriophage PBS2 has a DNA genome in which thymine is completely replaced by uracil attached to deoxyribose. It can survive in a host that carries a uracil-DNA glycosylase activity because the bacteriophage encodes the very strong peptide inhibitor Ugi[39] that binds irreversibly to the DNA-binding groove of the Ung-type uracil-DNA glycosylase of *B. subtilis*, as well as to the related DNA binding groove of mammalian UNG.[40] Furthermore, in another group of *B. subtilis* bacteriophages (e.g. phage 2C, SPO1 and Φe), 5-hydroxymethyluracil completely replaces thymine.[41] Although these are exceptions, they demonstrate that uracil and uracil analogues other than thymine can carry genetic information in a DNA organism, at least in small genomes like these bacteriophages.

2.5.1. *RNA Viruses have Small Genomes, High Mutation Rates and Inefficient Repair*

The RNA viruses are on average smaller than DNA viruses and encode relatively few genes. However, they encode their own RNA replicase that generates progeny viral RNA. RNA viruses frequently produce very high titres of progeny and display very high mutation rates ($\sim10^{-5}$ to 10^{-4} per site per replication) compared with DNA viruses ($\sim10^{-8}$ to 10^{-7}) in spite of the smaller genomes of RNA viruses.[32] Retroviruses also have very high mutation rates. The high mutation rate of RNA viruses and retroviruses is largely caused by the lack of proofreading by their RNA polymerases. Furthermore, the apparent lack of other effective mechanisms for RNA repair likely contributes to the very high mutation frequencies observed. Interestingly, there is a clear inverse relationship between genome size and mutation rate per base. The highest known mutation rate of any infectious RNA-agent known is that of the tiny viroids (10^{-3} to 10^{-2} per site per replication), the genomes of which are only ~250–460 nucleotides.[32] Some of the viroids have catalytic properties of ribozymes. This has led to the hypothesis that they may represent relics of an RNA World.[42] The high mutation rates of RNA viruses and retroviruses make them useful tools for studying evolution, but from a medical point of view it is problematic that the mutation rates may rapidly lead to drug-resistance.[43,44] Furthermore, the combined properties of high titers and high mutation rates may result in genetic diversity of progeny even in single cells, as demonstrated for vesicular stomatitis virus (VSV).[45] There is good evidence for direct repair of RNA methylations in both bacteriophage RNA[46] and bacterial RNA[47,48] by *E. coli* protein AlkB in live bacteria. Unfortunately, the physiological significance of RNA methylation repair remains unknown, but it does take place in endogenous RNA after physiological induction of AlkB.[48] Excision repair mechanisms require a complementary strand for "replacement repair," that in contrast to direct base repair, is a multistep process. Although many RNA viruses have dsRNA genomes and others may adopt RNA hairpin structures, there is to our knowledge no information on excision repair of viral RNA, neither by host factors or virus-encoded proteins.

While high mutation rates and high titers of progeny contribute to rapid evolution of RNA viruses, mutations may also reduce fitness, defined as ability to compete in an environment, including survival, reproduction and transmission.[43] Harmful mutations are probably more common than useful ones, therefore most mutations probably are not observed because the mutants cannot compete.

2.5.2. *Large DNA Viruses may Encode Complete Repair Pathways*

Several of the "semi-large" DNA viruses, like herpes virus and pox virus, encode some DNA repair enzymes, but may rely on host proteins to obtain the full complement of proteins in a pathway, depending on the replication status of the cell. Most of the viral DNA repair proteins are associated with BER, including uracil-DNA glycosylase, AP-endonucleases, ligases, PNK (polynucleotide kinase), FEN (flap endonuclease), PCNA-like molecules and helicases.[49] As an example, poxviruses encode PolX type polymerase (related to mammalian Polβ), AP-endonuclease (Nfo-like) and DNA ligase.[49] Poxvirus-encoded UNG was also found to be essential for DNA replication[50] and an obligate part of the replication machinery.[51,52] In addition, there is strong evidence suggesting that herpesvirus-encoded UNG enhances replication and replicative reactivation,[53-55] particularly in tissues expressing low levels of host UNG2.[53] Furthermore, herpes UNG also interacts directly with the viral replication complex.[55] Thus, pox- and herpesvirus UNG apparently have roles both in viral DNA repair and DNA replication.

Generally, viruses have no metabolism and cannot proliferate outside a host cell. They only become "alive" inside their host cells. However, the relatively recent discovery of the giant DNA mimivirus (mimicking microbes) and subsequently other giant viruses infecting amoebas in marine and fresh water environments has challenged both the definition of life and the history of evolution.[56] The *Acanthamoeba polyphaga* mimivirus encodes an Ung-type uracil-DNA glycosylase that, similar to mammalian UNG proteins, contains a long and unstructured N-terminal domain that may be required for structural and functional integrity. The crystal structure of the catalytic domain contains functional motifs that are closely related to those of other known UNG structures.[57] These apparently ancient giant viruses frequently have genomes larger than 1 million base pairs (1 Mbps) encoding more than 500 proteins,[35,58] including DNA repair proteins for complete pathways and genes required for transcription.[59] They are visible in the light microscope and were initially thought to be bacteria. The largest among giant viruses are in the family of *Pandoraviridae*, the genomes of which are ~2 Mbps encoding more than 2,000 proteins.[60,61]

The genomes of some giant viruses are larger than those of small intracellular bacteria. Thus, the genome of the smallest known bacterium, *Mycoplasma genitalium*, is only 0.58 Mbps and encodes 524 genes. Another small bacterium, *Mycoplasma gallisepticum*, has genes encoding proteins involved in several different repair pathways, including alkylation repair, mismatch repair (MMR), nucleotide excision repair (NER), base excision repair (BER), recombination and SOS repair.[62] Numerous other very small bacterial genomes have also been identified and sequenced.[63] Unlike the giant viruses, these small bacteria have energy metabolism and can survive outside of their host. This is possibly not an important distinction since genome reduction may have occurred during evolution of the giant viruses, as part of adaption to intracellular life. Interestingly, phylogenetic studies based on conservation of structures, rather than sequences, have led to the suggestion that giant viruses may represent a fourth and very ancient branch of life, in addition to Bacteria, Archaea and Eukarya.[35] However, it has been argued that the term "domain" should be restricted to descendants of the last universal common ancestor (LUCA) and that the basis for establishing giant viruses as a separate domain is not strong enough.[34]

2.6. Last Universal Common Ancestor (LUCA) and Transition from RNA to DNA Cells

It is thought that all forms of cellular life originate from a hypothetical LUCA. Accepting this view as a working hypothesis, a speculative chain of events leading to the three domains of life, Bacteria, Archaea and Eukarya, can be depicted. LUCA evolved into the present-day Bacteria along one branch, whereas the other branch later split into Archaea and Eukarya (Fig. 2.3). This view is to a large extent based upon the relationship between their respective ribosomal RNA. Thus, the conclusion is that Eukarya and Archaea are more related than Eukarya and Bacteria. It has been argued that since eukaryotes have an archaeal origin, Bacteria and Archaea represent the two primary domains of life.[64] Briefly, the transition from an RNA cell to a DNA cell allowed the evolution of larger and chemically more stable genomes, efficient genome maintenance processes and differentiation

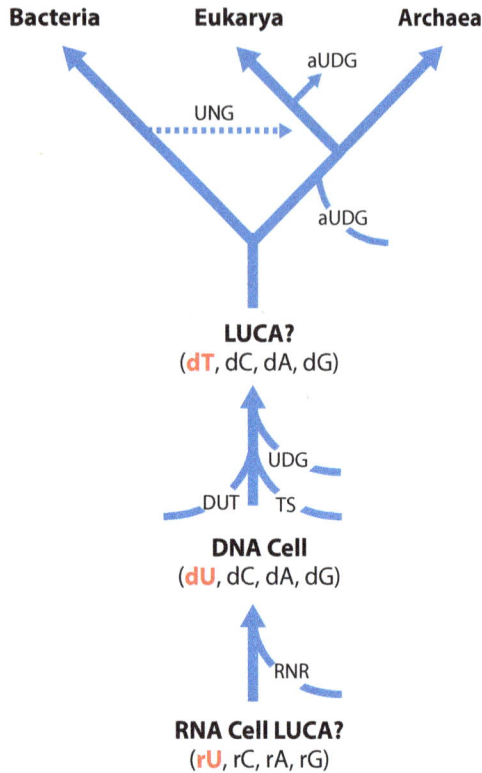

Fig. 2.3.   RNA cells and evolution of genetically stable DNA cells. The first cells may have been RNA cells with unstable genomes. It is unknown whether LUCA (last universal common ancestor) was an RNA cell or a DNA cell, although RNA cells most likely preceded DNA cells. The steps towards a DNA cell containing thymine may have included an intermediate DNA cell with genomic uracil resulting from evolution of ribonucleotide reductase (RNR), later to be replaced by DNA cells containing thymine. This required the evolution of a dUTPase (DUT gene) and thymidylate synthase (TS). Although this would result in a much more stable genome, aberrant genomic uracil still occurred due to incorporation of dUTP or deamination of cytosine. This problem was solved by evolution of uracil-DNA glycosylase (UDG), which may initially simply have been an enzyme that initiated degradation of faulty DNA.

into complex organisms. Central to this transition is the generation of deoxyribonucleotides from ribonucleotides and subsequently the replacement of uracil by thymine as carrier of genetic information.

2.6.1. *Were Primitive RNA Cells the First Form of Life?*

The first self-replicating molecules may have been RNA-like molecules that catalysed their own replication. Probably numerous different small RNA molecules were formed. As discussed above, the start was not necessarily a pure RNA world, but rather an RNA + peptide world. If tRNA-like molecules were generated, the 3' amino acid acceptor stem may have been loaded with specific amino acids by an "urzyme" synthase, according to a "hidden" code with a non-independent relationship to an anticodon like structure in the looped RNA. Thus, a primitive genetic code may have evolved early.[28,29] Concentration of RNA and other molecules within membrane-like structures may have facilitated catalysis eventually generating the first primitive RNA cells with metabolism. Possibly several different cell lineages competed, most of which disappeared.[33] An important factor in this process would be the evolution of a primitive ribosome ancestor using an RNA catalyst to form peptide bonds. This idea is supported by the finding that ribosomal RNA is extensively involved in all steps of peptide bond formation, suggesting that proteins may have arrived as integral parts of the ribosome later in evolution.[65]

2.6.2. *Was LUCA an RNA Cell or a DNA Cell?*

All known present-day Archaea and prokaryotes are DNA cells, but it has not been proven that LUCA was a DNA cell.[66] In fact, many investigators consider an RNA cell as a likely LUCA (Fig. 2.3). It may have been a primitive prokaryote, but some recent papers hypothesise that LUCA may actually have been a eukaryote-like RNA cell.[67-69] It has also been suggested that there might have been many LUCAs that at one stage had the ability to rapidly exchange genetic material, eventually evolving into the extant forms of life.[70,71] One would intuitively assume that LUCA or LUCAs were very simple cells with a minimal set of genes that could support life. As mentioned, the smallest known cellular genome, that of *M. genitalium*, encodes only 524 genes. However, LUCA may well have been considerably larger than this, since adaptation of such prokaryotes to intracellular life could have resulted in gene reduction.

2.6.3. *Transition From an RNA Cell to a DNA Cell in Steps*

DNA as genetic material has the important advantage that the DNA backbone is much more stable than that of RNA.[72-74] This would allow expansion of the genome size. Thus, the likely transition from a putative RNA cell to a DNA cell was obviously a major step in evolution and required the biosynthesis of dNTPs. The only known *de novo* pathway for generating deoxyribonucleotides is via ribonucleotides. The key enzyme is ribonucleotide reductase (RNR) that uses ribonucleotide diphosphates (rNDPs) as substrate in eukaryotes, but frequently rNTPs in prokaryotes, depending on the class of RNR present in the species.[75] RNRs are very complex metalloproteins that appear to have a common ancestor.[76] Simulation of an RNA-by-DNA genetic takeover suggested a central role of RNR even when varying a number of important chemical and environmental parameters.[77] If LUCA was indeed an RNA cell, RNR must have evolved in early descendants of LUCA before branching, or later by horizontal gene transfer. Alternatively, RNR may have evolved prior to LUCA, which would then presumably be a DNA cell.[66] As discussed, the first DNA cell most likely had uracil as a base instead of thymine in the genome. One argument supporting the U-DNA cell as an intermediate in the evolution is the fact that RNR can directly produce dATP, dGTP, dCTP and dUTP from the corresponding rNTPs, but RNR cannot directly generate dTTP. Instead, dUMP (from dUDP or dUTP) is used to generate dTMP by thymidylate synthase (TS or TYMS) and then dTTP in two steps by kinases.[75]

Considering the complexity of the catalysis by TS and the simultaneous requirement for a dUTPase, there must probably have been a time gap between the evolution of RNR and TS and hence a gap between the occurrence of the U-DNA cell and the T-DNA cells. However, the U-DNA cell would have a genome stability problem even in a short-term evolutionary perspective, particularly in organisms with large genomes, because of cytosine deamination to uracil and high mutation rates in the absence of DNA repair. This uracil would be chemically identical to genomic uracil from dUMP-incorporation, except that the deamination product would be present in a U:G mismatch, whereas incorporated uracil would be in a U:A pair. The genomic uracil-problem was solved by the evolution of three enzymes; dUTPase, thymidylate synthase (TS) and uracil-DNA glycosylase (UDG). The dUTPase gene (*dut*) is present in all domains of life, probably due to horizontal transfer of a gene of single origin.[78,79] The dUTPase strongly reduces dUTP levels and thereby prevent excessive incorporation of dUMP. The product, dUMP, was used by TS to generate dTMP, which was further phosphorylated to dTTP. UDGs would remove spurious genomic uracil resulting from cytosine deamination or dUMP-incorporation. It is thought that present-day types of RNR and UDG have single ancestor genes.[76,80] This was also thought to be the case for TS. However, it was found that the hyperthermophilic anaerobic archaeon *Pyrococcus abyssi*, which did not have the canonical TS of type ThyA (TYMS in mammalian cells), could still produce dTMP *de novo*. This was possible due to the presence of a different type of TS, called ThyX. This TS is deeply rooted in evolution and might actually be an ancestral form later replaced by the canonical ThyA type TS in many organisms. Surprisingly, ThyX is more widely distributed than ThyA in hyperthermophilic microorganisms and is present in several human pathogens as well.[81,82] Since it is possibly the most ancient form of TS, it was suggested that it could have been involved in the transition from a U-DNA world to a T-DNA world.[81]

2.7. Processing of Aberrant Genomic Uracil Takes Place in all Domains of Life

DNA repair as a mechanism may have evolved more than three billion years ago in an early prokaryote. The enzymes that today contribute to DNA repair may have had different functions early in evolution, e.g. in DNA replication and degradation. Once a repertoire of replication/repair enzymes had evolved, gene duplication and divergence could establish different and more specialized pathways.[83] DNA repair is present in all domains of life and so are uracil-DNA glycosylases (UDGs), the enzymes that initiate base excision repair (BER) of uracil in DNA. Genes for the established five families of UDGs apparently evolved from a common ancestral gene, perhaps already in the LUCA some 2.9 billion years ago.[84,85]

2.7.1. *The Uracil-DNA Glycosylase Family Comprises at Least Five Distinct Members*

A number of years after the seminal discovery of Ung[86] encoded by the *ung* gene, several other uracil-DNA glycosylases have been discovered. An updated overview of the UDG families identified by Lucas-Lledó *et al.* [85] is shown in Table 2.1 and in Fig. 2.4. Here extensive and robust identification of likely homologs of representative UDGs in complete proteomes (see table legend for details) has been used to map the distribution of UDGs across the main kingdoms of life. Both the general distribution of UDG families, which is discussed in more detail below, and the active site sequence logos are very similar to the original analysis by Lucas-Lledó *et al.*[85] The corresponding protein structures, where the conservation at the level of amino acid residues has been highlighted, show a strong conservation of the protein core and at the entrance of what is the uracil-binding pocket in UNG, and this conservation is particularly strong in UNG itself. However, the figure also illustrates the structural diversity of these proteins outside of the main core. This probably reflects the adaptive evolution of these proteins, where they have developed somewhat complementary functions. It has been shown for bacteria that loss of family 2 increases the preservation of family 1, whereas loss of family 1 enhances preservation of family 4, although it is also clear that the different UDGs can replace each other to some extent. Although sequence similarity between UDG-families is limited to a few important motifs and residues,[85] their structures are better conserved, particularly between UDGs of family 1 (UNG-type), family 2 (TDG-type) and family 3 (SMUG1-type). In contrast, overall structures of the extended family 4, including the putative family 6, are rather different, although

Table 2.1.   Phylogenomic Distribution of the Uracil-DNA Glycosylase Superfamily

Protein family	Eukaryota	Bacteria	Archaea	Viruses	Logo
Proteomes (total)	878	4,602	216	1712	
% species	100%	100%	100%	100%	
% gene family	12%	62%	3%	23%	
1 (UNG)	852	2,845	4	306	
% species	97%	62%	2%	18%	
% gene family	21%	71%	0%	8%	
2a (TDG)	631	485	1	90	
% species	72%	11%	0%	5%	
% gene family	52%	40%	0%	8%	
2b	12	989	17	59	
% species	1%	21%	8%	3%	
% gene family	1%	92%	2%	5%	
3a (SMUG1)	125	62	0	11	
% species	14%	1%	0%	1%	
% gene family	63%	31%	0%	6%	
3b	1	82	0	5	
% species	0%	2%	0%	0%	
% gene family	1%	93%	0%	6%	

(*Continued*)

Table 2.1.   (*Continued*)

Protein family	Eukaryota	Bacteria	Archaea	Viruses	Logo
4	1	2,454	118	152	
% species	0%	53%	55%	9%	
% gene family	0%	90%	4%	6%	
5	1	718	112	32	
% species	0%	16%	52%	2%	
% gene family	0%	83%	13%	4%	
6	0	799	0	78	
% species	0%	17%	0%	5%	
% gene family	0%	91%	0%	9%	

The table shows an updated distribution of protein families (1–6) originally identified by Lucas-Lledó *et al.*[85] The updated version was made by identifying representative protein sequences with known 3D structure for each family (see Fig. 2.4), and use these as query sequences in iterative PSI-Blast[88] searches of UniProtKB Reference Proteomes[89] at EBI[90], with an E-value cut-off at 1.0e-6 and at most 3 iterations. A known 3D structure could not be found for Family 3b, where W4TBP2_9FLAO was used as query sequence. Sequences representing significant hits were collected in libraries, and aligned using Muscle[91] in order to create logos using WebLogo.[92] For sequences with significant hits to more than one family the additional hits were removed, keeping the one with best E-value. The table shows the total number of proteomes in the database library, as well as the number of proteomes with significant hits for each family. The "% species" shows the percentage of significant hits relative to the total number of proteomes for each domain. The "% gene family" shows the percentage of significant hits relative to the total number of proteomes across all domains. Percentages >10 are highlighted in blue.

(A)

(B) (cont.)

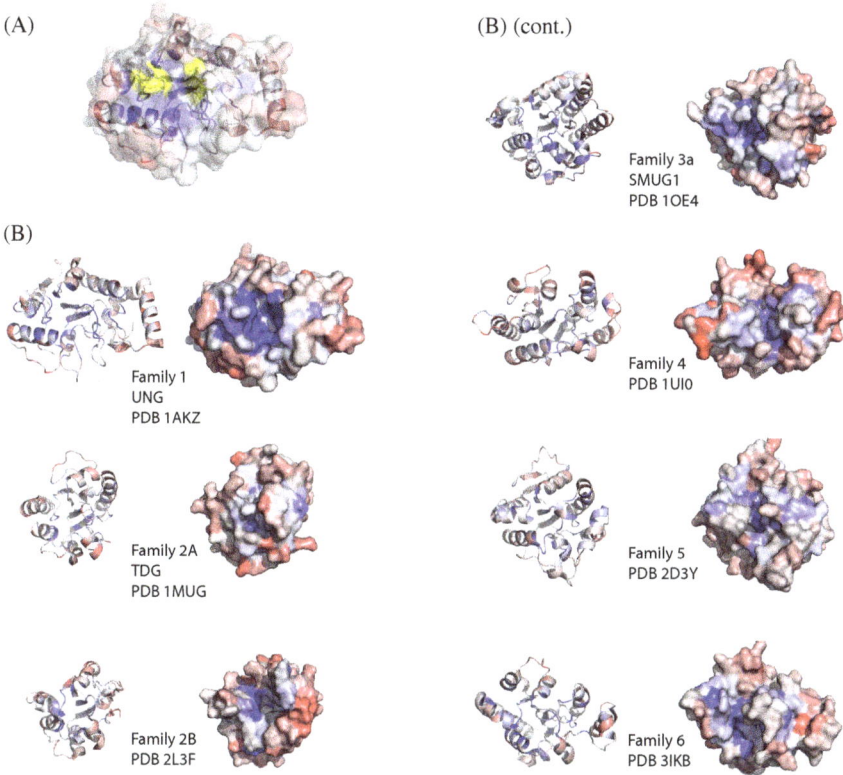

Family 3a
SMUG1
PDB 1OE4

(B)

Family 1
UNG
PDB 1AKZ

Family 4
PDB 1UI0

Family 2A
TDG
PDB 1MUG

Family 5
PDB 2D3Y

Family 2B
PDB 2L3F

Family 6
PDB 3IKB

Fig. 2.4.   Conservation of residues within protein families. The figure shows the pattern of amino acid conservation in the protein families listed in Table 2.1. The conservation was estimated using the Shannon score at each position of each multiple alignment described in Table 2.1, and visualized on a red-white-blue colour scale from highly variable (red) to strongly conserved (blue) on a cartoon representation of each structure, and a surface representation of the same structure. The structures were drawn with PyMOL (The PyMOL Molecular Graphics System, Version 1.8 Schrödinger, LLC). (A) Combined cartoon/surface representation of the reference structure, PDB entry 1AKZ. Here the residues listed as critical to enzyme activity by[103] are highlighted in yellow. (B) Visualization of each family, except Family 3b. In each case the protein structure was structurally aligned to the reference structure, using the PyMOL cealign command.

their active sites have some resemblance to that of families 1–3.[85] Sequence conservation of proteins within families is often extensive at the amino acid level. For example, the catalytic domain of *E. coli* Ung and the human orthologue, UNG, display ~55% amino acid identity and ~75% conservation. At the time of cloning, this was the closest relationship between any human and *E. coli* protein known.[87]

This high degree of conservation may suggest horizontal gene transfer from a prokaryote as an origin of at least some UDGs in complex organisms. As one example, the eukaryotic *UNG* gene is probably acquired by horizontal transfer of a bacterial gene (Fig. 2.3).

UDGs evolved into the now distantly related enzymes with somewhat different properties. UNG-enzymes are the most commonly present UDG, but not universally present.[85] Both prokaryotic and eukaryotic UNG-proteins are highly selective for uracil, have very high catalytic turnover numbers and remove uracil efficiently from U:A pairs, U:G mismatches and single stranded DNA.[93,94] TDGs/ Mugs are also found in many prokaryotes and eukaryotes. They have a broader substrate range, are essentially double strand- and mismatch-specific, but have very low turnover numbers.[95,96] SMUG1 (single strand selective monofunctional uracil-DNA glycosylase 1)[97] is most common in eukaryotes, but is also found in some prokaryotes.[85,98] Similar to TDG/Mug, SMUG1 has somewhat different properties in different organisms, but prefers double stranded DNA in mammalian cells. The full name of SMUG1 may thus be considered somewhat misleading. SMUG1 proteins have low/intermediate turnover numbers and have some preference for mismatches.[94] Archaea have none of these proteins. Instead, they have a form of UDG that is, at best, very distantly related to other UDGs, even structurally. In addition, Archaea have polymerases that discriminate against dUTP incorporation. Thus, different organisms use different strategies to eliminate or prevent the presence of genomic uracil. However, no known cellular organism lacks mechanisms to avoid genomic uracil. It is intriguing that that eukaryotes and Archaea use very different enzymes to remove or avoid genomic uracil, considering that eukaryotic cells are thought to have branched off from an early form of Archaea. The much stronger similarity between bacterial and mammalian UNGs, as compared with other vertebrate UDGs may suggest that mammalian UNG is a result of horizontal gene transfer and that an ancient archaeal form of UDG may have been lost in evolution. At some stage in vertebrate development, UNG-type UDG acquired a non-canonical role in adaptive immunity. Adaptive immunity, a process involving somatic hypermutation (SHM), class switch recombination (CSR) and/or gene conversion, evolved in several genetic steps.[99] Thus, UNG and AID are required for gene conversion in chicken cells[100] and for SHM and CSR in mammals.[101] Furthermore, non-templated SHM takes place in shark immunoglobulin light chain genes.[102] Although the enzymatic mechanism was not examined in the cited paper, the observed SHM indicate that the role of AID and UNG in adaptive immunity may be very old, perhaps as much as 400 million years.

2.8. Some Perspectives on Genomic Uracil

All domains of life, as well as DNA viruses, have evolved mechanisms to remove genomic uracil. Nevertheless, small to moderate increases in genomic uracil are apparently well tolerated even in mammals, as exemplified by mouse models defective in a single uracil-DNA glycosylase (UNG or SMUG1) or two major UDGs (UNG *and* SMUG1). TDG-deficiency is embryonic lethal, most probably because this protein is also a transcription factor, and has a central function in epigenetics as well. MMR may be a last resort to remove genomic uracil, but even triple knockout mice deficient in UNG, SMUG1 and MSH2 are viable and fertile, although they have a shortened life span and develop lymphomas.[104] In spite of the viability and fertility of organisms lacking uracil-removal systems, the universal presence of such systems in cells, and even many viruses, indicates that long-term survival of a species is not possible without mechanisms to remove uracil from the genome. Furthermore, in the immune system genomic uracil has taken on novel protective functions, rather than destructive, during evolution. These protective functions generally play on the regulated conversion of cytosine to uracil to obtain altered DNA sequences with advantageous properties. One major question is whether the immune system is the only example of this duality of genomic uracil.

References

1. Leu, K., Obermayer, B., Rajamani, S., *et al.* The prebiotic evolutionary advantage of transferring genetic information from RNA to DNA. *Nucleic Acids Res* **39**, 8135–8147 (2011).

2. Zahnle, K., Schaefer, L.,, Fegley, B. Earth's earliest atmospheres. *Cold Spring Harb Perspect Biol* **2**, a004895 (2010).

3. Eme, L., Sharpe, S. C., Brown, M. W., Roger, A. J. On the age of eukaryotes: evaluating evidence from fossils and molecular clocks. *Cold Spring Harb Perspect Biol* **6** (2014).

4. Knoll, A. H. Paleobiological perspectives on early eukaryotic evolution. *Cold Spring Harb Perspect Biol* **6** (2014).

5. Poole, A. M.,, Gribaldo, S. Eukaryotic origins: How and when was the mitochondrion acquired? *Cold Spring Harb Perspect Biol* **6**, a015990 (2014).

6. Jeltsch, A. Oxygen, epigenetic signaling, and the evolution of early life. *Trends Biochem Sci* **38**, 172–176 (2013).

7. Ruiz-Trillo, I., Burger, G., Holland, P. W., *et al.* The origins of multicellularity: a multi-taxon genome initiative. *Trends Genet* **23**, 113–118 (2007).

8. Morris, S. C., Caron, J. B. A primitive fish from the Cambrian of North America. *Nature* **512**, 419–422 (2014).

9. Shu, D. G., Luo, H. L., Conway Morris, S., *et al.* Lower Cambrian vertebrates from south China. *Nature* **402**, 42–46 (1999).

10. Killian, J. K., Buckley, T. R., Stewart, N., *et al.* Marsupials and Eutherians reunited: genetic evidence for the Theria hypothesis of mammalian evolution. *Mamm Genome* **12**, 513–517 (2001).

11. Shoshani, J., McKenna, M. C. Higher taxonomic relationships among extant mammals based on morphology, with selected comparisons of results from molecular data. *Mol Phylogenet Evol* **9**, 572–584 (1998).

12. Chimpanzee Sequencing and Analysis Consortium. Initial sequence of the chimpanzee genome and comparison with the human genome. *Nature* **437**, 69–87 (2005).

13. Franchini, L. F., Pollard, K. S. Genomic approaches to studying human-specific developmental traits. *Development* **142**, 3100–3112 (2015).

14. Fu, Q., Hajdinjak, M., Moldovan, O. T., *et al.* An early modern human from Romania with a recent Neanderthal ancestor. *Nature* **524**, 216–219 (2015).
15. Paabo, S. The human condition-a molecular approach. *Cell* **157**, 216–226 (2014).
16. Kuhlwilm, M., Gronau, I., Hubisz, M. J., *et al.* Ancient gene flow from early modern humans into Eastern Neanderthals. *Nature* **530**, 429–433 (2016).
17. Pizzarello, S., Shock, E. The organic composition of carbonaceous meteorites: the evolutionary story ahead of biochemistry. *Cold Spring Harb Perspect Biol* **2**, a002105 (2010).
18. Robertson, M. P., Joyce, G. F. The origins of the RNA world. *Cold Spring Harb Perspect Biol* **4** (2012).
19. Higgs, P. G., Lehman, N. The RNA World: molecular cooperation at the origins of life. *Nat Rev Genet* **16**, 7–17 (2015).
20. Horning, D. P., Joyce, G. F. Amplification of RNA by an RNA polymerase ribozyme. *Proc Natl Acad Sci U S A* **113**, 9786–9791 (2016).
21. Poudyal, R. R., Nguyen, P. D., Lokugamage, M. P., *et al.* Nucleobase modification by an RNA enzyme. *Nucleic Acids Res* (2016).
22. Prywes, N., Blain, J. C., Del Frate, F., Szostak, J. W. Nonenzymatic copying of RNA templates containing all four letters is catalyzed by activated oligonucleotides. *Elife* **5** (2016).
23. Shapiro, R. Prebiotic cytosine synthesis: a critical analysis and implications for the origin of life. *Proc Natl Acad Sci U S A* **96**, 4396–4401 (1999).
24. Saladino, R., Botta, G., Delfino, M., Di Mauro, E. Meteorites as catalysts for prebiotic chemistry. *Chemistry (Easton)* **19**, 16916–16922 (2013).
25. Saladino, R., Carota, E., Botta, G., *et al.* Meteorite-catalyzed syntheses of nucleosides and of other prebiotic compounds from formamide under proton irradiation. *Proc Natl Acad Sci U S A* **112**, E2746–2755 (2015).
26. Sponer, J. E., Sponer, J., Novakova, O., *et al.* Emergence of the First Catalytic Oligonucleotides in a Formamide-Based Origin Scenario. *Chemistry (Easton)* **22**, 3572–3586 (2016).
27. Ferus, M., Nesvorny, D., Sponer, J., *et al.* High-energy chemistry of formamide: a unified mechanism of nucleobase formation. *Proc Natl Acad Sci U S A* **112**, 657–662 (2015).
28. Carter, C. W., Jr., Wolfenden, R. tRNA acceptor stem and anticodon bases form independent codes related to protein folding. *Proc Natl Acad Sci U S A* **112**, 7489–7494 (2015).
29. Wolfenden, R., Lewis, C. A., Jr., Yuan, Y., Carter, C. W., Jr. Temperature dependence of amino acid hydrophobicities. *Proc Natl Acad Sci U S A* **112**, 7484–7488 (2015).

30. Shapiro, R. A replicator was not involved in the origin of life. *IUBMB Life* **49**, 173–176 (2000).

31. Holmes, E. C. Molecular clocks and the puzzle of RNA virus origins. *J Virol* **77**, 3893–3897 (2003).

32. Holmes, E. C. What does virus evolution tell us about virus origins? *J Virol* **85**, 5247–5251 (2011).

33. Forterre, P. The two ages of the RNA world, and the transition to the DNA world: a story of viruses and cells. *Biochimie* **87**, 793–803 (2005).

34. Forterre, P., Krupovic, M., Prangishvili, D. Cellular domains and viral lineages. *Trends Microbiol* **22**, 554–558 (2014).

35. Nasir, A., Caetano-Anolles, G. A phylogenomic data-driven exploration of viral origins and evolution. *Sci Adv* **1**, e1500527 (2015).

36. Nasir, A., Forterre, P., Kim, K. M., Caetano-Anolles, G. The distribution and impact of viral lineages in domains of life. *Front Microbiol* **5**, 194 (2014).

37. Mager, D. L., Stoye, J. P. Mammalian Endogenous Retroviruses. *Microbiol Spectr* **3**, MDNA3-0009-2014 (2015).

38. Cherkasova, E., Weisman, Q., Childs, R. W. Endogenous retroviruses as targets for antitumor immunity in renal cell cancer and other tumors. *Front Oncol* **3**, 243 (2013).

39. Friedberg, E. C., Ganesan, A. K., Minton, K. N-Glycosidase activity in extracts of Bacillus subtilis and its inhibition after infection with bacteriophage PBS2. *J Virol* **16**, 315–321 (1975).

40. Mol, C. D., Arvai, A. S., Sanderson, R. J., *et al.* Crystal structure of human uracil-DNA glycosylase in complex with a protein inhibitor: protein mimicry of DNA. *Cell* **82**, 701–708 (1995).

41. Hoet, P. P., Coene, M. M., Cocito, C. G. Replication cycle of Bacillus subtilis hydroxymethyluracil-containing phages. *Annu Rev Microbiol* **46**, 95–116 (1992).

42. Diener, T. O. Circular RNAs: relics of precellular evolution? *Proc Natl Acad Sci U S A* **86**, 9370–9374 (1989).

43. Hughes, D., Andersson, D. I. Evolutionary consequences of drug resistance: shared principles across diverse targets and organisms. *Nat Rev Genet* **16**, 459–471 (2015).

44. Vermehren, J., Sarrazin, C. The role of resistance in HCV treatment. *Best Pract Res Clin Gastroenterol* **26**, 487–503 (2012).

45. Combe, M., Garijo, R., Geller, R., Cuevas, J. M., Sanjuan, R. Single-Cell Analysis of RNA Virus Infection Identifies Multiple Genetically Diverse Viral Genomes within Single Infectious Units. *Cell Host Microbe* **18**, 424–432 (2015).

46. Aas, P. A., Otterlei, M., Falnes, P. O., Vagbo, C. B., Skorpen, F., Akbari, M., Sundheim, O., Bjoras, M., Slupphaug, G., Seeberg, E. *et al.* Human and bacterial oxidative demethylases repair alkylation damage in both RNA and DNA. *Nature* **421**, 859–863 (2003).

47. Ougland, R., Zhang, C. M., Liiv, A., Johansen, R. F., Seeberg, E., Hou, Y. M., Remme, J., Falnes, P. O. AlkB restores the biological function of mRNA and tRNA inactivated by chemical methylation. *Mol Cell* **16**, 107–116 (2004).

48. Vagbo, C. B., Svaasand, E. K., Aas, P. A., Krokan, H. E. Methylation damage to RNA induced in vivo in Escherichia coli is repaired by endogenous AlkB as part of the adaptive response. *DNA Repair (Amst)* **12**, 188–195 (2013).

49. Redrejo-Rodriguez, M., Salas, M. L. Repair of base damage and genome maintenance in the nucleo-cytoplasmic large DNA viruses. *Virus Res* **179**, 12–25 (2014).

50. Millns, A. K., Carpenter, M. S., DeLange, A. M. The vaccinia virus-encoded uracil DNA glycosylase has an essential role in viral DNA replication. *Virology* **198**, 504–513 (1994).

51. Burmeister, W. P., Tarbouriech, N., Fender, P., *et al.* Crystal Structure of the Vaccinia Virus Uracil-DNA Glycosylase in Complex with DNA. *J Biol Chem* **290**, 17923–17934 (2015).

52. Holzer, G. W., Falkner, F. G. Construction of a vaccinia virus deficient in the essential DNA repair enzyme uracil DNA glycosylase by a complementing cell line. *J Virol* **71**, 4997–5002 (1997).

53. Minkah, N., Macaluso, M., Oldenburg, D. G., *et al.* T. Absence of the uracil DNA glycosylase of murine gammaherpesvirus 68 impairs replication and delays the establishment of latency in vivo. *J Virol* **89**, 3366–3379 (2015).

54. Pyles, R. B., Thompson, R. L. Evidence that the herpes simplex virus type 1 uracil DNA glycosylase is required for efficient viral replication and latency in the murine nervous system. *J Virol* **68**, 4963–4972 (1994).

55. Su, M. T., Liu, I. H., Wu, C. W., *et al.* Uracil DNA glycosylase BKRF3 contributes to Epstein-Barr virus DNA replication through physical interactions with proteins in viral DNA replication complex. *J Virol* **88**, 8883–8899 (2014).

56. Raoult, D., Audic, S., Robert, C., *et al.* The 1.2-megabase genome sequence of Mimivirus. *Science* **306**, 1344–1350 (2004).

57. Kwon, E., Pathak, D., Chang, H. W., Kim, D. Y. Crystal structure of mimivirus uracil-DNA glycosylase. *PLoS One* **12**, e0182382 (2017).

58. Sharma, V., Colson, P., Pontarotti, P., Raoult, D. Mimivirus inaugurated in the 21st century the beginning of a reclassification of viruses. *Curr Opin Microbiol* **31**, 16–24 (2016).

59. Fischer, M. G., Kelly, I., Foster, L. J., Suttle, C. A. The virion of Cafeteria roenbergensis virus (CroV) contains a complex suite of proteins for transcription and DNA repair. *Virology* **466–467**, 82–94 (2014).

60. Abergel, C., Legendre, M., Claverie, J. M. The rapidly expanding universe of giant viruses: Mimivirus, Pandoravirus, Pithovirus and Mollivirus. *FEMS Microbiol Rev* **39**, 779–796 (2015).

61. Scheid, P. A strange endocytobiont revealed as largest virus. *Curr Opin Microbiol* **31**, 58–62 (2016).

62. Gorbachev, A. Y., Fisunov, G. Y., Izraelson, M., *et al*. DNA repair in Mycoplasma gallisepticum. *BMC Genomics* **14**, 726 (2013).

63. McCutcheon, J. P. The bacterial essence of tiny symbiont genomes. *Curr Opin Microbiol* **13**, 73–78 (2010).

64. Williams, T. A., Foster, P. G., Cox, C. J., Embley, T. M. An archaeal origin of eukaryotes supports only two primary domains of life. *Nature* **504**, 231–236 (2013).

65. Moore, P. B., Steitz, T. A. The involvement of RNA in ribosome function. *Nature* **418**, 229–235 (2002).

66. Poole, A. M., Logan, D. T. Modern mRNA proofreading and repair: clues that the last universal common ancestor possessed an RNA genome? *Mol Biol Evol* **22**, 1444–1455 (2005).

67. Forterre, P. The universal tree of life: an update. *Front Microbiol* **6**, 717 (2015).

68. Glansdorff, N., Xu, Y., Labedan, B. The last universal common ancestor: emergence, constitution and genetic legacy of an elusive forerunner. *Biol Direct* **3**, 29 (2008).

69. Mariscal, C., Doolittle, W. F. Eukaryotes first: how could that be? *Philos Trans R Soc Lond B Biol Sci* **370**, 20140322 (2015).

70. Koonin, E. V. Comparative genomics, minimal gene-sets and the last universal common ancestor. *Nat Rev Microbiol* **1**, 127–136 (2003).

71. Koonin, E. V. On the origin of cells and viruses: primordial virus world scenario. *Ann N Y Acad Sci* **1178**, 47–64 (2009).

72. Lindahl, T. Irreversible heat inactivation of transfer ribonucleic acids. *J Biol Chem* **242**, 1970–1973 (1967).

73. Lindahl, T. Instability and decay of the primary structure of DNA. *Nature* **362**, 709–715 (1993).

74. Singer, B., Fraenkel-Conrat, H. The Nature of the Breaks Occurring in Tmv-Rna under Various Conditions. *Biochim Biophys Acta* **76**, 143–145 (1963).

75. Hofer, A., Crona, M., Logan, D. T., Sjoberg, B. M. DNA building blocks: keeping control of manufacture. *Crit Rev Biochem Mol Biol* **47**, 50–63 (2012).

76. Lundin, D., Berggren, G., Logan, D. T., Sjoberg, B. M. The origin and evolution of ribonucleotide reduction. *Life (Basel)* **5**, 604–636 (2015).

77. Ma, W., Yu, C., Zhang, W., Wu, S., Feng, Y. The emergence of DNA in the RNA world: an in silico simulation study of genetic takeover. *BMC Evol Biol* **15**, 272 (2015).

78. McClure, M. A. Evolution of the DUT gene: horizontal transfer between host and pathogen in all three domains of life. *Curr Protein Pept Sci* **2**, 313–324 (2001).

79. Baldo, A. M., McClure, M. A. Evolution and horizontal transfer of dUTPase-encoding genes in viruses and their hosts. *J Virol* **73**, 7710–7721 (1999).

80. Aravind, L., Koonin, E. V. The alpha/beta fold uracil DNA glycosylases: a common origin with diverse fates. *Genome Biol* **1**, RESEARCH0007 (2000).

81. Leduc, D., Graziani, S., Meslet-Cladiere, L., *et al.* Two distinct pathways for thymidylate (dTMP) synthesis in (hyper)thermophilic Bacteria and Archaea. *Biochem Soc Trans* **32**, 231–235 (2004).

82. Myllykallio, H., Lipowski, G., Leduc, D., *et al.* An alternative flavin-dependent mechanism for thymidylate synthesis. *Science* **297**, 105–107 (2002).

83. O'Brien, P. J. Catalytic promiscuity and the divergent evolution of DNA repair enzymes. *Chem Rev* **106**, 720–752 (2006).

84. Nasir, A., Kim, K. M., Caetano-Anolles, G. Giant viruses coexisted with the cellular ancestors and represent a distinct supergroup along with superkingdoms Archaea, Bacteria and Eukarya. *BMC Evol Biol* **12**, 156 (2012).

85. Lucas-Lledo, J. I., Maddamsetti, R., Lynch, M. Phylogenomic analysis of the uracil-DNA glycosylase superfamily. *Mol Biol Evol* **28**, 1307–1317 (2011).

86. Lindahl, T. An N-glycosidase from Escherichia coli that releases free uracil from DNA containing deaminated cytosine residues. *Proc Natl Acad Sci USA* **71**, 3649–3653 (1974).

87. Olsen, L. C., Aasland, R., Wittwer, C. U., *et al.* Molecular cloning of human uracil-DNA glycosylase, a highly conserved DNA repair enzyme. *EMBO J* **8**, 3121–3125 (1989).

88. Altschul, S. F., Madden, T. L., Schaffer, A. A., *et al.* Gapped BLAST and PSI-BLAST: a new generation of protein database search programs. *Nucleic Acids Res* **25**, 3389–3402 (1997).

89. The UniProt Consortium. UniProt: the universal protein knowledgebase. *Nucleic Acids Res* **45**, D158–D169 (2017).
90. Li, W., Cowley, A., Uludag, M., *et al.* The EMBL-EBI bioinformatics web and programmatic tools framework. *Nucleic Acids Res* **43**, W580–584 (2015).
91. Edgar, R. C. MUSCLE: multiple sequence alignment with high accuracy and high throughput. *Nucleic Acids Res* **32**, 1792–1797 (2004).
92. Crooks, G. E., Hon, G., Chandonia, J. M., Brenner, S. E. WebLogo: a sequence logo generator. *Genome Res* **14**, 1188–1190 (2004).
93. Kavli, B., Sundheim, O., Akbari, M., *et al.* hUNG2 is the major repair enzyme for removal of uracil from U:A matches, U:G mismatches, and U in single-stranded DNA, with hSMUG1 as a broad specificity backup. *J Biol Chem* **277**, 39926–39936 (2002).
94. Krokan, H. E., Drablos, F., Slupphaug, G. Uracil in DNA--occurrence, consequences and repair. *Oncogene* **21**, 8935–8948 (2002).
95. Cortazar, D., Kunz, C., Saito, Y., *et al.* The enigmatic thymine DNA glycosylase. *DNA Repair (Amst)* **6**, 489–504 (2007).
96. Dong, L., Mi, R., Glass, R. A., *et al.* Repair of deaminated base damage by Schizosaccharomyces pombe thymine DNA glycosylase. *DNA Repair (Amst)* **7**, 1962–1972 (2008).
97. Haushalter, K. A., Todd Stukenberg, M. W., Kirschner, M. W., Verdine, G. L. Identification of a new uracil-DNA glycosylase family by expression cloning using synthetic inhibitors. *Curr Biol* **9**, 174–185 (1999).
98. Pettersen, H. S., Sundheim, O., Gilljam, K. M., *et al.* Uracil-DNA glycosylases SMUG1 and UNG2 coordinate the initial steps of base excision repair by distinct mechanisms. *Nucleic Acids Res* **35**, 3879–3892 (2007).
99. Flajnik, M. F., Kasahara, M. Origin and evolution of the adaptive immune system: genetic events and selective pressures. *Nat Rev Genet* **11**, 47–59 (2010).
100. Arakawa, H., Buerstedde, J. M. Activation-induced cytidine deaminase-mediated hypermutation in the DT40 cell line. *Philos Trans R Soc Lond B Biol Sci* **364**, 639–644 (2009).
101. Di Noia, J. M., Neuberger, M. S. Molecular mechanisms of antibody somatic hypermutation. *Annu Rev Biochem* **76**, 1–22 (2007).
102. Lee, S. S., Tranchina, D., Ohta, Y., Flajnik, *et al.* Hypermutation in shark immunoglobulin light chain genes results in contiguous substitutions. *Immunity* **16**, 571–582 (2002).

103. Parikh, S. S., Mol, C. D., Slupphaug, G., *et al.* Base excision repair initiation revealed by crystal structures and binding kinetics of human uracil-DNA glycosylase with DNA. *EMBO J* **17**, 5214–5226 (1998).

104. Kemmerich, K., Dingler, F. A., Rada, C., Neuberger, M. S. Germline ablation of SMUG1 DNA glycosylase causes loss of 5-hydroxymethyluracil- and UNG-backup uracil-excision activities and increases cancer predisposition of Ung-/-Msh2-/- mice. *Nucleic Acids Res* **40**, 6016–6025 (2012).

Chapter 3

Routes to Uracil in DNA

Geir Slupphaug*, Bodil Kavli and Hans E. Krokan

Uracil in DNA results from incorporation of dUMP instead of dTMP by DNA polymerases during replication, or by spontaneous or enzymatic deamination of already existing DNA cytosines. Which of these that quantitatively dominates in a given cell depends on many factors including cell type and proliferative status, genetic background and exposure to e.g. genotoxic or infectious agents. This chapter will provide an overview of the different routes to genomic uracil, with a major focus on mammalian cells.

3.1. Incorporation of Uracil During Replication

3.1.1. *Utilization of dUTP and dTTP as Building Blocks by DNA Polymerases*

In 1958, the Kornberg laboratory demonstrated that a partially purified DNA polymerase fraction from *E. coli* was able to utilise dUTP, as well as 5-bromo-dUTP, instead of dTTP in enzymatic synthesis of DNA.[1] Since cellular kinases that could phosphorylate dUMP to dUTP were not known, the authors concluded that uracil was likely not incorporated in DNA *in vivo*. This was in line with the general view in the 1950s, that natural DNA contained no uracil.[2,3] Three

* geir.slupphaug@ntnu.no

years later, the group of Peter Reichard at Karolinska Institutet discovered a ribonucleoside diphosphate reductase (RNR) in *E. coli* that converts CDP to dCDP.[4] Shortly thereafter they demonstrated that the purified enzyme fraction also catalyzed conversion of UDP to dUDP and (less efficiently) UTP to dUTP, thus demonstrating a potential route to dUTP formation *in vivo*.[4] They also identified an activity in the same fraction that was able to convert dUTP to dUMP and pyrophosphate, which marked the discovery of dUTPase.[4] From these findings they concluded, *"The absence of uracil in DNA may thus not only be due to the lack of an enzyme which phosphorylates deoxy-UMP to deoxyUTP but also to the presence of an enzyme which specifically cleaves deoxyUTP."* Parallel studies by Greenberg and Somerville at the University of Michigan demonstrated that infection of *E. coli* by bacteriophage T2 led to expression of a kinase system able to convert dUMP to dUDP and dUTP.[5] They also confirmed the existence of a strong dUTPase activity in *E. coli*. Nevertheless, they proposed that under certain conditions such as in thymineless- or dUTPase mutants, uracil could actually be present in DNA below the levels of detection attainable at the time. DNA bases were commonly quantified subsequently to acid hydrolysis of isolated DNA and separation by thin-layer chromatography, and the authors roughly estimated that about 100 uracil residues per bacterial genome could go undetected using this method.[5] In the following year it was found that in DNA isolated from *B. subtilis* infected with the transducing phage PBS2, dTMP was completely replaced with dUMP.[6] This definitely demonstrated that uracil-containing DNA could be formed *in vivo*. It was suggested that phage-mediated rewiring of the bacterial pyrimidine metabolism could be mediating this, as two enzymes were exclusively found in the infected cells that could affect the levels of dUTP and dTTP; a dUMP kinase that phosphorylated dUMP and a dTMPase that dephosphorylated dTMP and prevented formation of dTTP.[7] It was later demonstrated, however, that these enzymes were of minor importance and that expression of a recently discovered enzyme, dCTP deaminase, was the major mediator of dUTP and uracilated phage DNA.[8] Shortly thereafter, dCTPase was also identified in *Salmonella typhimurium*[9] and in *E. coli*.[10]

Even though these studies provided strong evidence for alternative pathways for synthesis of cellular dUTP, the general view still prevailed that incorporation of dUMP during replication in normal cells was negligible. In fact, the discovery of *E. coli* uracil-DNA glycosylase by Tomas Lindahl in 1974[11] even substantiated this, by adding a second enzyme (in addition to dUTPase) that was able to prevent permanent inclusion of uracil in DNA. However, in the following year Konrad and Lehman identified strains of *E. coli* with a mutation in the *sof* (*dnaS*) locus, which could incorporate label within short DNA fragments following [³H]thymidine pulse labelling.[12] These fragments were much smaller than Okazaki fragments, but were nevertheless transformed into high molecular weight DNA. It was thus speculated whether these fragments could be intermediates in DNA replication. Tye and coworkers in the group of Bernhard Weiss, who provided evidence that the *dnaS* locus was identical to the *dut* locus encoding dUTPase, resolved this issue.[13] The dUTPase deficiency mediated incorporation of dUMP into DNA and subsequent base excision repair that mediated generation of transient short DNA fragments. The authors further suggested that at least a portion of the Okazaki fragments seen in wild type cells could be generated via this mechanism. Shortly thereafter, several groups went on to study the effects of dUTP during polyoma and animal cell replication, and obtained similar results as in *E. coli*.[14-16] Erik Wist and the group of Hans Krokan in Tromsø demonstrated that isolated nuclei of HeLa cells were able to incorporate [³H]dUMP during replication and that most of the incorporated radioactivity was rapidly removed from the endogenous DNA during further incubation. Moreover, the presence of dUTP generated very small DNA fragments that eventually became ligated. Thus, their work probably represents the first observation of mammalian DNA base excision repair.[15] They also demonstrated that a partially purified DNA polymerase α from HeLa nuclei incorporated [³H]dUMP and [³H]dTMP with equal efficiency, supporting the previous observations in *E. coli* by the Kornberg laboratory.

Subsequent work supported that many bacterial and eukaryotic DNA polymerases utilized dUTP and dTTP as building blocks for DNA replication with similar efficiency. Partially purified DNA polymerases α and β from calf thymus[17] and human cells as well as DNA polymerase β from rat neurons[18,19] had similar K_M values for dUTP and dTTP. On the other hand, some selectivity for dTTP over dUTP was observed among the high-fidelity nuclear and mitochondrial DNA polymerases. DNA polymerase γ from porcine liver had about threefold lover K_M for dTTP (0.4 μM) than for dUTP (1.1 μM)[20] and DNA polymerase δ from HeLa cells had twofold lower K_M for dTTP (1 μM) than for dUTP (2 μM).[19] Human DNA polymerase γ, however, had essentially identical K_M values for dUTP (0.149 μM) and dTTP (0.140 μM).[21] In summary, these studies indicated that the incorporation of dUMP into bacterial and eukaryotic DNA would be largely dependent on the relative cellular pool sizes of dUTP and dTTP.

3.1.2. *Enzymatic Regulation of Cellular dUTP and dTTP Pools*

Cellular ribonucleotide and deoxyribonucleotide pools are in a constant state of flux, and are subject to multiple levels of regulation, including allosteric and genetic controls.[22] A major overall upregulation of dNTP synthesis occurs in S-phase, where a eukaryotic cell incorporates about 4×10^6 nucleotide/sec into the replicating DNA. This is so rapid that the entire pool of dNTPs can be depleted within one minute after initiation of replication unless rapidly replenished. Thus, in S-phase dNTP pool increases to a level 20-fold higher than in G1 phase to support copying of the entire genome.[23] The mammalian pyrimidine metabolism is illustrated in Fig. 3.1 and has been the topic of several reviews (see e.g. Ref. 24). The following section will thus primarily focus on the enzymatic steps that are most relevant for regulation of the dUTP/dTTP ratio.

3.1.3. *Synthesis of dUTP*

Ribonucleotide reductase

Central to the formation of dUTP and dTTP as well as the other deoxyribonucleotides is ribonucleotide reductase (RNR). RNR

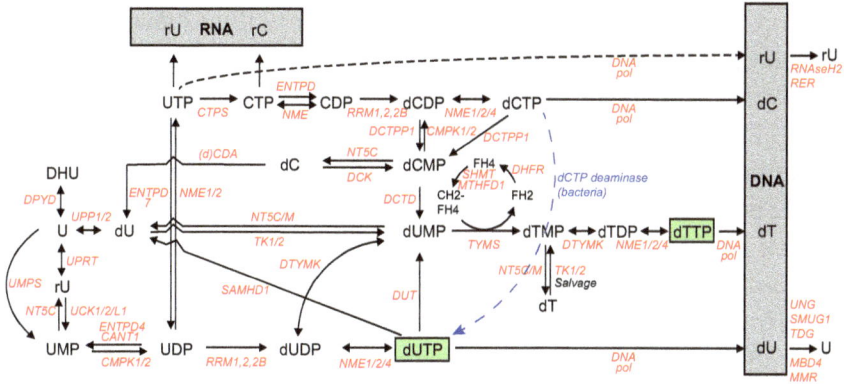

Fig. 3.1. Overview of pyrimidine metabolism. Enzymes involved in the metabolism are in italics. CANT1 (soluble calcium-activated nucleotidase 1); (d)CDA [(deoxy) cytidine deaminase]; CH2-FH4 (5,10-methylenetetrahydrofolate; CMPK (UMP-CMP kinase); CTPS (CTP synthase); DCK (deoxycytidine kinase); DCTD (deoxycytidylate deaminase); DCTPP1 (dCTP pyrophosphatase 1); DHFR (dihydrofolate reductase); DPYD (dihydropyrimidine dehydrogenase); DTYMK (thymidylate kinase); DUT (deoxyuridine 5'-triphosphate nucleotidohydrolase); ENTPD (ectonucleoside triphosphate diphosphohydrolase); FH2 (dihydrofolate); FH4 (tetrahydrofolate); NME (nucleoside diphosphate kinase); MTHFD1 (methylenetetrahydrofolate dehydrogenase); NT5C/M (5'-deoxyribonucleotidase, cytosolic/mitochondrial); RRM1, 2, 2B [ribonucleoside-diphosphate reductase (RNR), subunits 1, 2, 2B]; SAMHD1 (deoxynucleoside triphosphate triphosphohydrolase); SHMT (serine hydroxymethyltransferase); TK (thymidine kinase); TYMS (thymidylate synthase); UCK (uridine-cytidine kinase); UMPS (uridine 5'-monophosphate synthase); UPP (uridine phosphorylase); UPRT (uracil phosphoribosyltransferase homolog).

converts UDP, CDP, ADP and GDP to their corresponding dNTPs by direct reduction.[25] dTDP is instead formed by conversion of dUMP to dTMP by thymidylate synthase and further phosphorylation to dTDP and dTTP (Fig. 3.1). RNR is often denoted the gatekeeper of dNTP homeostasis[26] and constitutes the rate-limiting step in *de novo* dUTP synthesis. The enzyme is a tetramer consisting of two copies of the large catalytic (RRM1) and two copies of either RRM2 or the more recently discovered RRM2B subunit.[27,28] RNR activity is subject to tight regulation via several mechanisms including allosteric and oligomeric regulation as well as alteration of the levels and subcellular localization of the RNR subunits. Allosteric regulation occurs via binding of either dATP/ADP, dTTP or dGTP to the catalytic subunit, and binding of dATP/ATP induces UDP/CDP reduction. However, the overall activity of the enzyme is inhibited at high dATP/ATP ratios (Ref. 29 and references therein). The catalytic RRM1 subunit is expressed throughout the cell cycle, but the smaller RRM2 and RRM2B subunits are differentially regulated. Whereas RRM2 peaks in S-phase, RRM2B is evenly expressed throughout the cell cycle. Moreover, after DNA damage, RRM2B is induced in a p53-dependent manner and is thus often denoted p53R2. Together with RRM1, RRM2B forms an active holoenzyme that fuels *de novo* synthesis of dNTPs for DNA repair in quiescent cells.[27,30] The RRM1/2B complex is also present in mitochondria, where RRM2B is involved in ribonucleotide reduction as well as in thioredoxin reduction, which is needed for DNA synthesis and repair.[31]

Nucleoside diphosphate kinases

These kinases (NME1/2/4) catalyze interconversion of dUDP and dUTP (as well as other di- and triphosphonucleosides)[32] using ATP as the major phosphate donor. Thus, at the high ATP-concentrations in the cytosol, the reaction likely favours dUTP formation. The eukaryotic NDKs belong to the NM23 (non-metastatic 23) gene family, which is highly conserved between eubacteria, archaea and eukaryotes. In humans, nine genes (NME1-9) encode such enzymes, among which NME1-4 and 6 have been shown to possess (d)NDP kinase activity, although they may also act as exonucleases and protein kinases and thus regulate a multitude of cellular processes, including DNA repair ([33] and references therein).

3.1.4. Synthesis of dTTP

Whereas RRN and NME are major players in the formation of dUTP, formation of dTTP is somewhat more complicated. In the *de novo* synthetic route, dUMP is a key intermediate that can be formed by at least three enzymatic steps.

Deoxyuridine 5'-triphosphate nucleotidohydrolase (dUTPase)

dUTPase irreversibly dephosphorylates dUTP to dUMP and inorganic pyrophosphate[5] and thus serves as a crucial regulator the dUTP/dTTP ratio by concomitantly contributing to reduce the dUTP- and increase the dTTP levels. (reviewed in[34]) This dual role also explains why the enzyme is essential in many prokaryotes and eukaryotes.[35] However, in a dUTPase knockout background, viability can be restored in some species by simultaneous knockout of uracil-DNA glycosylase activity.[36,37] In fact, in one of the first site-directed mutagenesis protocols, Thomas A. Kunkel used an *E. coli* strain lacking both dUTPase and UNG activity.[38] Recent findings indicate that the activities of RNR and dUTPase must be balanced to avoid accumulation of cellular dUTP. When the cell-cycle regulated RRM2 subunit of RNR was overexpressed in a non-tumorigenic human epithelial cell line, this led to elevated cellular dUTP as well as increased genomic uracil, replication stress and strand breaks at AT-rich common fragile sites (CFS). These effects were, however, reversed by concomitant overexpression of dUTPase.[39] Since CFS are replicated late in S-phase, the authors suggest that depletion of dTTP in late S leads to increased incorporation of dUMP at these AT-rich sequences and BER-mediated replication stress. Interestingly, a combination of low dUTPase and high RNR levels in tumours is correlated with poor prognosis in colorectal- and breast cancer, which could aid therapeutic decision-making.[39]

Interestingly, in *Drosophila melanogaster*, dUTPase expression is stage- and tissue-specific, and down-regulated during the larval stages. This downregulation is inversely correlated with the genomic uracil level, which may amount to 200–2,000 per million bases during the larval stages[40] and is comparable to the levels observed in dUTPase- and Ung-deficient *E. coli*.[41] Such high levels of genomic uracil are probably tolerated because *D. melanogaster* lacks the uracil-DNA glycosylase UNG. However, tolerance is restricted to the larval stage, and the authors propose that the transient downregulation of dUTPase and a tissue-specific loss of genomic uracil tolerance at the post-larval stage are essential for fruit fly metamorphosis and probably for other Holometabola insects. In some organisms, like *Mycobacteria* and *Plasmodia*, the dUTPase activity constitutes the only known route to dUMP and dTTP synthesis, and dUTPase inhibitors have thus become an attractive intervention to fight malaria and tuberculosis.[34] Many non-primate retroviruses also encode their own dUTPase, which has recently been highlighted as a potential target to combat retroviral diseases.[42]

Deoxycytidylate deaminase (DCTD)

dUMP can also be formed by deamination of dCMP by deoxycytidylate deaminase. This enzyme has been found in a wide range of species, including humans.[43] DCTD has recently received renewed interest since it can lead to inactivation of the monophosphorylated forms of deoxycytidine analogues that are used in anti-cancer and antiviral therapies.[44] It is allosterically regulated, activated by dCTP and inhibited by the end-product of the pathway, dTTP.[45]

Thymidine kinase (TK1/2)

TK1 and TK2 can mediate both *de novo* and salvage synthesis of dTTP, by utilising either dU or dT as substrates, respectively (Fig. 3.1). Both the cytosolic, S-phase specific TK1 and the mitochondrial, constitutively expressed TK2 readily accept both dT and dU as substrates, although the enzymes have about 20-fold lower K_M for dT than for dU. ([46] and references therein) Since the steady state level of dU is low in many normal tissues such as heart, kidney, liver, muscle and peripheral nerves,[47] *de novo* synthesis of dTTP from dU is likely of less significance than from dCTP and dUTP.

Thymidylate synthase reaction

Disregarding the source of dUMP, a crucial step in the biosynthesis of dTTP is methylation of dUMP to dTMP, which is catalyzed by thymidylate synthase (TYMS).[48] In the reaction, 5,10-methylenetetrahydrofolate (CH2FH4, 5,10-CH2THF) (Fig. 3.1), derived from the essential micronutrient folate, donates the methyl group to dUMP to form dTMP and dihydrofolate (FH2, DHF). To allow continuous catalysis, CH2FH4 must be regenerated via two enzymatic steps, first to tetrahydrofolate (FH4, THF) by dihydrofolate reductase (DHFR) and then to CH2FH4 by serine hydroxymethyltransferase (SHMT). In the DHFR reaction, NADPH acts as electron donor, and in the SHMT reaction, the source of the methyl carbon is L-serine that is converted to glycine. Remethylation of FH4 is catalyzed by methylenetetrahydrofolate dehydrogenase (MTHFD1). The thymidylate synthesis reaction has been the theme of several comprehensive reviews,[49,50] which also highlight the reaction as target of many cytostatic and antimicrobial drugs. Inhibition of the reaction either by targeting TYMS or the CH2FH4-regeneration steps concomitantly reduces dTTP synthesis and increases dUTP accumulation and thus has a profound effect on the cellular dUTP/dTTP ratio.

Thymidylate kinase (DTYMK)

DTYMK catalyses the phosphorylation of dTMP derived either from thymidylate synthase or from salvage synthesis by TK1, to form dTDP,[51,52] and is thus the first common step of both pathways (Fig. 3.1). Inhibition of DTYMK would thus block dTTP synthesis (via further phosphorylation by NME1/2/4) from both the *de novo* and the salvage pathway. DTYMK has received renewed interest as a potential cancer drug target following findings that knockdown of DTYMK sensitises cancer cells to low doses of agents that induce DNA double strand breaks (DSBs) and repair by homologous recombination (HR).[53] Since the other dNDPs are formed by RNR, and many cancer cells have an elevated ratio of RRM2/DTYMK, the authors proposed that that these cancer cells are selectively dependent on DTYMK to produce sufficient dTTP to facilitate synthesis of the long stretches of DNA formed during HR. Knockdown or inhibition of DTYMK would then lead to incorporation of dUTP and cytotoxic futile repair cycles.[53] Likely, this should mediate fewer side effects than inhibition of thymidylate synthase, which would have lower specificity for transformed cells.

3.1.5. *Degradation of dUTP and dTTP*

In addition to the dUTPase step and the reversible NME-catalysed dephosphorylation steps of dUTP and dTTP described above, the more recently discovered deoxynucleoside triphosphate triphosphohydrolase (SAMHD1) is able to irreversibly dephosphorylate all the dNTPs to their respective deoxyribonucleosides.[54] SAMHD1 is induced in response to viral infection[55] and apparently acts as a lentiviral restriction factor by lowering the concentration of dNTPs in myeloid cells to levels that fail to support reverse transcription.[56] Apparently, SAMHD1 does not discriminate against dUTP as a substrate compared to the canonical dNTPs.[57] Thus, yet undiscovered mechanisms must contribute to maintaining the dUTP levels high when the canonical dNTPs are depleted, causing a high dUTP/dTTP ratio in resting immune cells.[58] SAMHD1 is, however, subject to complex regulation[57] and a mutant form (R366H) of the enzyme found in colon adenocarcinoma has been shown to have altered substrate preference compared to wild-type.[59] This mutant had reduced catalytic activity against dCTP, dTTP and dATP, whereas dGTP hydrolysis was unaffected. It is also possible that specific post-translational modifications or regulatory factors may affect the substrate preference of SAMHD1. Interestingly, recent work has shown that specific phosphorylations and acetylations in SAMDH1 markedly affect its viral restriction[60] and catalytic activity.[61-63] However, in these studies the activity towards dUTP relative to the canonical dNTPs was not addressed.

3.1.6. *Quantification of Cellular dUTP and dTTP Pools*

Cellular deoxyribonucleotides have proven challenging to quantify due to their low concentrations (pmol/10^6 cells) and the large excess (100-1,000-fold) of the corresponding ribonucleotides. A commonly employed method to detect and quantify various cellular nucleobases prior to the evolution of HPLC and mass spectrometry methods was (two-dimensional) paper- or thin-layer chromatography. However, early versions of these methods could achieve only little or no separation between dUTP and dTTP.[64,65] This was resolved in 1978 by employing polyethyleneimine-coated cellulose plates, an acidic solvent in the first dimension to resolve uracil from thymine compounds, and salt elution in the second dimension to separate by charge.[66] This method allowed the first detection and quantification of dUTP *in vivo* levels and demonstrated that subsequent to infection of *B. subtilis* by the bacteriophage PBS1, the cellular dTTP level dropped sharply and the corresponding dUTP level increased gradually during the first 20 min after infection, simultaneously with the onset of DNA synthesis. This confirmed the prediction that dUTP was the probable substrate for replication of the uracil-containing PBS1 genome.[66]

To quantify cellular dUTP in eukaryotic cells in the presence of much higher levels of dTTP, Nilsson and co-workers devised an indirect enzymatic method in which dUTP/dTTP in cell extracts were incorporated into a poly(dA:dT) substrate by *E. coli* DNA polymerase Klenow fragment together with labelled dATP. After excision of uracil from the product by UNG, alkaline treatment and sucrose gradient size fractionation allowed estimation of the relative amounts of dUTP and dTTP, when dUTPase-treated extracts were used as controls.[67] This method allowed determination of dUTP concentrations when these were between 0.1% and 2% of the dTTP pool. Whereas dUTP in uninfected mouse fibroblasts was below the detection limit, the cellular dUTP/dTTP ratio increased subsequent to polyoma virus infection and was found to be about 0.004 at the peak of DNA synthesis.[67] Importantly, the Klenow subunit of *E. coli* DNA polymerase I employed in similar polymerase assays by several groups, has been found to incorporate some NTPs in addition to dNTPs. Current assays have thus replaced the enzyme with Thermo Sequenase, which discriminates well between NTPs and dNTPs.[68]

By employing a sensitive, but laborious method including [³H] dU- labelling, periodate conversion of ribonucleotides, anion chromatography, dephosphorylation of dUTP to dU, paper chromatography and HPLC, Goulian *et al.* [69] found the level of dUTP in untreated human lymphoblasts to be ≤ 0.3 fmol/10^6 cells (≤ 0.3 nM). In the same cells, the dTTP level was found to be ~40 pmol/10^6 cells, giving a dUTP/dTTP ratio of $\leq 10^{-5}$. Treatment of the cells with methotrexate to inhibit thymidylate synthase (TYMS) leads to a dramatic increase in dUTP, which was easily detected at ~0.2 pmol/10^6 cells, with a concomitant drop in dTTP to ~1 pmol/10^6 cells. The increased dUTP:dTTP ratio (~0.2) suggested that substantial amounts of dUMP were incorporated during DNA replication in the drug-treated cells, and that the resulting fragmentation of uracil-containing DNA could contribute to the cytotoxicity.[69] The complexity of this method and similar multistep methods[70] that require quite large amounts of starting material (10^7 cells or more) and laborious sample handling, spurred the development of alternative assays, including a radioimmunoassay with sensitivity in the low fmol range and requiring less than 10^6 cells.[71] However, this assay required removal of dUMP and dUTP prior to the antibody reaction and failed to give precise quantification of dUTP in untreated cells. In 1997, Horowitz *et al.* devised a modified DNA-polymerase based assay capable of quantifying subpicomol levels of dUTP in standard solutions.[72] Although less sensitive than the method employed by Goulian *et al.*, it was adopted by several groups to monitor altered dUTP/dTTP ratios in human cancer cell lines subsequent to TYMS inhibition or knockdown of dUTPase.[73]

By coupling liquid chromatography (LC) to mass spectrometry (MS/MS) the requirement for pre-depletion of ribonucleotides from cell extracts was largely alleviated.[79] By employing this method, an unexpectedly high dUTP/dTTP ratio of 60 was found in terminally differentiated macrophages (Table 3.1.).[58] Notably, a recent method combined separation of mono-, di- and triphosphate moieties by strong anion exchange, with dephosphorylation of target fractions to molar equivalent nucleosides, prior to quantification by LC-MS/MS. This method had a sensitivity in the fmol range per 10^6 for the four canonical dNTPs, but unfortunately dUTP was not included in the analyses.[80]

1976, Shin *et al.* found that about 10% of total folates (cofactors in the enzymatic conversion of dUMP to dTMP) in liver resided in the nuclear compartment,[85] suggesting that folate metabolism could occur in the nucleus. Prem veer Reddy and Pardee proposed the existence of a "replitase" that synthesized dNTPs at the replication fork in S-phase.[86] In the following years, they conducted a series of experiments supporting the existence of such a complex in mammals, and that suggested that the dNTPs produced within this nuclear complex were synthesized from NDPs and more readily channelled directly into DNA compared to surrounding free dNTPs. They also proposed that RNR, TK1, DTYMK, DHFR, NME and DNA polymerase α resided within the complex.[87–90] Another group concomitantly reported the purification of a multienzyme complex from human lymphoblastoid cells, which contained at least DNA polymerase, TK1, dTMP kinase, NME and TYMS.[91] When either [³H]thymidine or [³H]dTMP was added, the isolated complex channelled the precursors directly into DNA with little accumulation of free [³H]dTTP, suggesting that the DNA polymerase regulated the flow of intermediates through the complex. More recently, the idea of co-localization of dNTP synthesis and DNA replication has been supported by immunocytochemical analyses demonstrating that TYMS, DHFR and SHMT1/2 (and likely other enzymes) co-localise with sites of replication initiation and form a multienzyme complex. Here, the complex associates with replication proteins like PCNA and is tethered to the nuclear lamina via SHMT1/2[92] (Fig. 3.2). Notably, if such a complex also contains active RNR and NME it would facilitate local synthesis of both dUTP and dTTP at replication forks. To what extent the "replitase" complex mediates specific suppression of dUTP over dTTP synthesis has, to the best of out knowledge, not been addressed. Apparently, the individual enzymes within the complex may allosterically interact with each other[88] and thus comprise a potential mechanism to fine-tune the dUTP/dTTP ratio, e.g. by direct formation of dUMP from dUDP by DTYMK, from dU by TK1 or from dCMP by DCD (Fig. 3.2).

Among these, DTYMK and TK1 have been proposed to reside in the replitase complex. Alternatively, the presence of DUT would allow

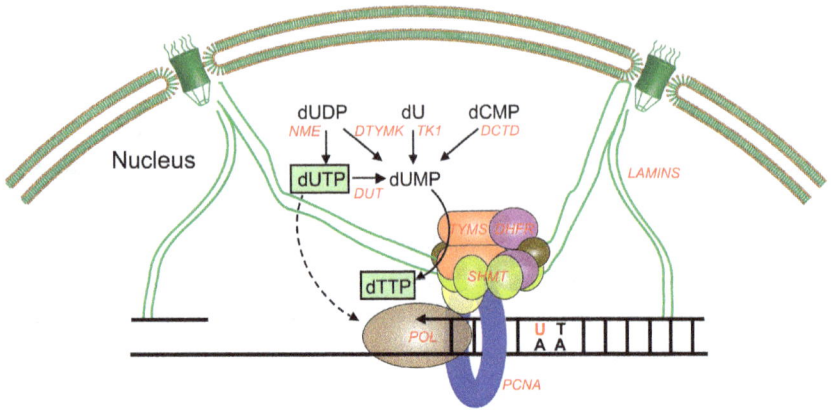

Fig. 3.2. Central proteins in thymidylate synthesis form a putative multienzyme «replitase» complex at sites of DNA replication (only one arm of replication fork illustrated). DHFR (dihydrofolate reductase); DTYMK (thymidylate kinase); DUT (deoxyuridine 5'-triphosphate nucleotidohydrolase); MTHFD1 (methylene-tetrahydrofolate dehydrogenase); NME (nucleoside diphosphate kinase); SHMT (serine hydroxymethyltransferase); TYMS (thymidylate synthase).

effective conversion of dUTP to dUMP and local suppression of the dUTP/dTTP ratio. Currently there is little evidence of co-localization of DUT with replication foci. However, a recent large scale study involving biochemical fractionation and mass spectrometry of samples from several species in fact indicates that this might occur, since DUT was found to co-fractionate with TYMS in *D. melanogaster*, *Xenopus laevis* and humans.[93] Whether DUT actually functionally interacts with the replication machinery and the replitase to locally regulate the dUTP/dTTP ratio, or other routes to dUMP are employed that bypass dUTP, remains to be established.

Regulation at DNA repair sites

The question of local regulation of the dUTP/dTTP ratio or preferential physical funnelling of dTTP over dUTP into the DNA polymerase, also applies to DNA repair synthesis. Several enzymes involved in pyrimidine metabolism have been demonstrated to localise at sites of DNA repair. This may be of special importance in quiescent and non-S-phase cells, where the overall dNTP pool is low and the dUTP/dTTP ratio is high. This should favor a mechanism involving locally increased dNTP synthesis and local suppression of the dUTP/dTTP ratio to avoid excessive misincorporation of dUMP and futile repair cycles. Several lines of evidence indicate that local production of dNTPs also occurs at DNA repair sites. RNR[94] and DTYMK[53] are both recruited to repair foci and produce dUDP (+ dCDP, dGDP and dADP) and dTDP, respectively. Recruitment of RNR to repair sites occurs via the RRM1 subunit, which can bind physically to the histone acetyltransferase Tip60 (KAT5)[94] and potentially contribute to co-ordinated dNTP synthesis, chromatin decondensation and DNA repair synthesis. NME1 is also recruited and ensures conversion of the deoxyribonucleoside diphosphates to triphosphates.[95] Whether a multienzyme complex similar to the replitase proposed at DNA replication sites is also recruited to repair sites has, however, not been established. Nevertheless, SHMT1 that is believed to serve a scaffolding function of the complex at replication sites, can also interact with several proteins involved in DNA repair. Notably, a distinct set of repair proteins was found to bind SHMT1 after UV-stress in non-replicating cells compared to those bound during S-phase.[92] Moreover, the three core components of the complex, TYMS, SHMT and DHFR all contain a consensus SUMOylation motif, and for SHMT, SUMOylation was shown to mediate nuclear translocation of SHMT in S-phase, that persisted until G2/M.[96] SUMOylation has been detected in many proteins involved in DNA repair, and has been proposed to constitute a central component of the DNA damage response.[97] Although circumstantial, the above findings provide some evidence that the complex might be recruited to repair foci outside of S-phase. Virtually no information exists regarding potential localization of DUT to DNA repair foci to suppress the local dUTP concentration. However, in a large scale study involving biochemical fractionation and mass spectrometry of samples from several species, DUT was found to co-fractionate with TYMS in *D. melanogaster*, *X. laevis* and humans.[93] Undoubtedly, further studies are warranted to clarify a potential role of local fine-tuning of the dUTP/dTTP ratio during DNA repair synthesis in non-proliferating cells.

Regulation in mitochondria

In mammalian cells mitochondrial DNA usually constitutes less than 1% of the nuclear DNA[98] and the mitochondrial dNTP pool represents only a small fraction of the total dNTP pool.[99] However, whereas chromosomal replication in nuclei only occurs in the S-phase, mitochondrial replication also occurs outside of S-phase. Thus, mitochondria adjust their dNTP pools independently of the overall cellular dNTP pools. The mitochondrial dNTP pools are localized in the mitochondrial matrix and are separated from the cytosol by a double lipid bilayer containing carrier proteins that are able to import dNTPs from the cytosol. Of these SLC25A33 imports dUTP, dTTP, and dCTP, as well as their corresponding (d)NDPs, whereas SLC25A36 imports dUTP and dCTP as well as the corresponding (d)NMPs and (d)NDPs.[100] Apparently, import of dNTPs from the cytosol is sufficient to sustain mitochondrial DNA replication in proliferating cells.

In postmitotic tissues and quiescent cells, when the cytosolic dNTP pool is low, mitochondria synthesize their own dNTPs. This has been believed to primarily occur through the salvage pathway ([101] and references therein) and the rate limiting enzyme for pyrimidine nucleotides in this pathway is the mitochondrial thymidine kinase (TK2).[102] TK2 is localized in the matrix[103] and converts thymidine imported from the cytosol, to dTMP. Further conversion of dTMP to dTDP is somewhat less clear. Some authors have proposed that this step is catalyzed by the cytosolic DTYMK,[104] which would require mitochondrial export/import of the substrate and product. A thymidylate kinase activity has, however, later been demonstrated in isolated mitochondria from rat heart[105] and mouse liver,[106] and a human gene (Loc129607) encoding a putative mitochondrial thymidylate kinase (named TMPK2) was identified in 2008.[107] The authors failed, however, to demonstrate any enzyme activity of the recombinant enzyme or in cell extracts overexpressing TMPK2, thus its substrate preference remains elusive. The final phosphorylation to dTTP is catalyzed by mitochondrial NME4.[108] The two final phosphorylation steps are readily reversible, so that dephosphorylation of dTMP to thymidine by the mitochondrial 5'-deoxyribonucleotidase (NT5M), can shift the equilibrium towards dephosphorylation of dTTP. Interestingly, recent findings indicate that the salvage synthesis of dTTP in mitochondria might be compartmentalized to prefer dTMP formed by TK2 over free (added) dTTP. When [³H]thymidine or [³H]TMP was added to isolated intact mitochondria from the rat heart, [³H]thymidine was readily metabolized to [³H]TTP, whereas far less [³H]TTP was produced with [³H]TMP. When mitochondrial membranes were broken by freezing/thawing, added [³H]thymidine was converted to [³H]TMP, but not to [³H]TTP.[109] The authors propose that the thymidylate kinase may not be functional unless the dTMP substrate is presented by TK2. Such a handover mechanism would also explain the lack of thymidylate synthase activity in the recombinant TMPK2 enzyme reported by Chen *et al.*[107] To what extent such compartmentalization also involves direct physical funnelling of dTTP into the mitochondrial replication machinery and thus discrimination against dUTP, is presently unknown.

The view that the salvage pathway was the primary (sole) synthetic route to dTTP within the mitochondria was challenged in 2011, when Anderson *et al.* [110] demonstrated the presence of an active *de novo* dTMP synthesis pathway in human mitochondria. This included TYMS, SHMT2 and DHFRL1. The finding that SHMT2 cross-links to mtDNA upon formaldehyde treatment[111] makes it tempting to speculate whether a multienzyme dTTP synthesis complex similar to that observed in nuclei, also exists at mitochondrial replication forks or at sites of mitochondrial DNA repair. Also, the potential presence of a dUTPase activity associated with the mitochondrial replication machinery remains elusive. Mammalian mitochondria contain a distinct isoform of DUT that is constitutively expressed,[112] which conforms well to the existence of active *de novo* mitochondrial dTTP synthesis.

3.1.8. *Do High dUTP/dTTP Ratios Provide Protection Against Viral Infection in Quiescent Cells?*

An obvious question related to the above findings is why some quiescent cells have extremely high overall dUTP/dTTP ratios? A potential explanation to this is that a high relative dUTP concentration would promote excessive incorporation of dUMP into the genomes of invading viruses and breakdown of the viral DNA via UDG-mediated BER. It is likely that the high overall dUTP/dTTP ratio in terminally differentiated macrophages[58] constitutes such a protective mechanism. Macrophages are prime targets for infection by retroviruses such as HIV-1. Since the HIV-1 reverse transcriptase, like most DNA polymerases, does not discriminate significantly between dUTP and dTTP, the genomes of viral progeny will be heavily uracilated. HIV reverse transcripts generated in primary human immune cells have been shown to contain >500 uracils per 10 kb HIV genome.[113] Surprisingly, the authors found that the high uracil content actually promoted the early life cycle of the virus, apparently by inhibiting suicidal autointegration and thus favoring chromosomal integration and viral replication. Other studies indicate, however, that the fate of the uracilated DNA depends on the amount of UNG2 present in the infected cells and that high nuclear UNG2 levels rather promote excessive BER-initiated strand breaks and degradation of the viral cDNA.[58,114] The notion that excessive uracilation via incorporation from the dNTP pool may negatively affect retroviral proliferation, is also supported by the fact that many primate and non-primate retroviruses encode their own dUTPases. (reviewed in [42]) Potentially, a similar protective function of a high dUTP/dTTP ratio could exist in several non-proliferating tissues and thus constitute a hitherto unappreciated part of the innate immunity towards invading DNA viruses. This question must, however, await future studies.

3.2. Deamination of DNA Cytosine

Whereas misincorporation of dUMP instead of dTTP during replication does not mediate a direct mutagenic effect, uracil generated by deamination of DNA cytosine will be 100% mutagenic unless repaired before the next round of replication and generate G:C to A:T transitions. Such deaminations thus pose a significant threat to the genomic integrity. Deamination of DNA cytosine can occur spontaneously under physiological conditions and the rate can be significantly enhanced by heat as well as some chemicals. In addition, nature has devised a mechanism to utilize enzymatically induced cytosine deamination to mediate intended mutagenic events as well as to counteract viral infection. In the following section, current knowledge of the different routes to cytosine deamination will be summarized.

3.2.1. *Spontaneous Deamination of DNA Cytosine*

In 1958, Schuster and Schramm reported that the amino groups of adenine, cytosine and guanine of the nucleic acids were deaminated by nitrous acid.[115] Parallel studies by Hurst and Kuskis demonstrated that hot alkali treatment deamination of DNA cytosine to uracil.[116] In search of alternative chemical methods of deamination and to develop potential mutagenic agents with higher specificity, Shapiro and Klein found that cytosine and cytidine were selectively deaminated in in acidic buffer solutions at pH up to 5.0.[117] Although the reaction was slow (time to half reaction was 16 h in 2 M citrate buffer, pH 4 at 95°C), they proposed that deamination of cytosine, rather than depurination, was the underlying cause of mutations previously observed in bacteriophage T4 subsequent to incubation in weakly acidic buffers at elevated temperatures (37–54°C).[118] Notably, in the same paper, the authors report that a small amount of deaminated cytidine was also produced by heating in distilled water or acetate buffer, pH 7.2.[117] Shortly thereafter, Tomas Lindahl reported that when the four ribonucleotides were incubated at 90°C in neutral buffers of physiological ionic strength, the primary degradation reaction was deamination of CMP to UMP.[119] In 1974, Lindahl and Nyberg published comprehensive study on heat-induced deamination of cytosine in single- and double-stranded *E. coli* DNA[120] and found the rate for deamination of C in single-stranded DNA at pH 7.4 and 97°C was $k = 2 \times 10^{-7}$ sec^{-1}. Importantly, spontaneous deamination of the base and nucleoside was found to be comparable to that of single-stranded DNA, whereas double-stranded DNA was deaminated 2–3 orders of magnitude more slowly.[120,121] Apparently, however, the rate of deamination in double-stranded DNA varies significantly depending on the local sequence context, since the rate of cytosine deamination in poly(dG)•poly(dC) resembled that of single-stranded DNA.[120] Based on their findings, Lindahl and Nyberg concluded that *"-cellular repair mechanisms may well exist that specifically convert guanine-uracil base pairs in DNA back to guanine-cytosine pairs"*. Two months later, they indeed published the identification of an N-glycosidase that released free uracil from DNA containing deaminated cytosines.[11]

The large difference between rates for single- and double-stranded DNA, and the lack of information on the fraction of DNA that on average is single-stranded, makes it difficult to predict exactly how many DNA-cytosines are deaminated per human genome per day. However, from available data, it has been calculated that if all DNA is double-stranded, 180 cytosines will be deaminated to uracil per day, whereas if all DNA were single-stranded the number would be 44,000. Assuming that 0.1% single-stranded DNA is a reasonable estimate for a human genome, then some 224 DNA-cytosines would be deaminated per day.[122] In these calculations, the possible influence of chromatin proteins was not considered. However, it would seem that spontaneous deamination of cytosine would be a quantitatively minor source of genomic uracil, but nevertheless important, since it generates highly mutagenic U:G mismatches.

Two mechanisms of cytosine deamination under normal physiological conditions have been suggested: (i) a direct, acid-catalyzed route involving protonation at the N3 position and attack by H_2O of the bond between C4 and the amino group, followed by release of NH_4^+, or (ii) an addition-elimination mechanism involving the addition of H_2O to the 5,6-double bond of protonated C3 and release of NH_4^+ and elimination of H_2O. Of these, the direct, acid-catalyzed route is illustrated Fig. 3.3.

3.2.2. Nitrosative Deamination of DNA Cytosine

Deamination of DNA cytosine may also be induced by peroxynitrite (N_2O_3, ion: $ONOO^-$),[123,124] which can be formed either via reaction

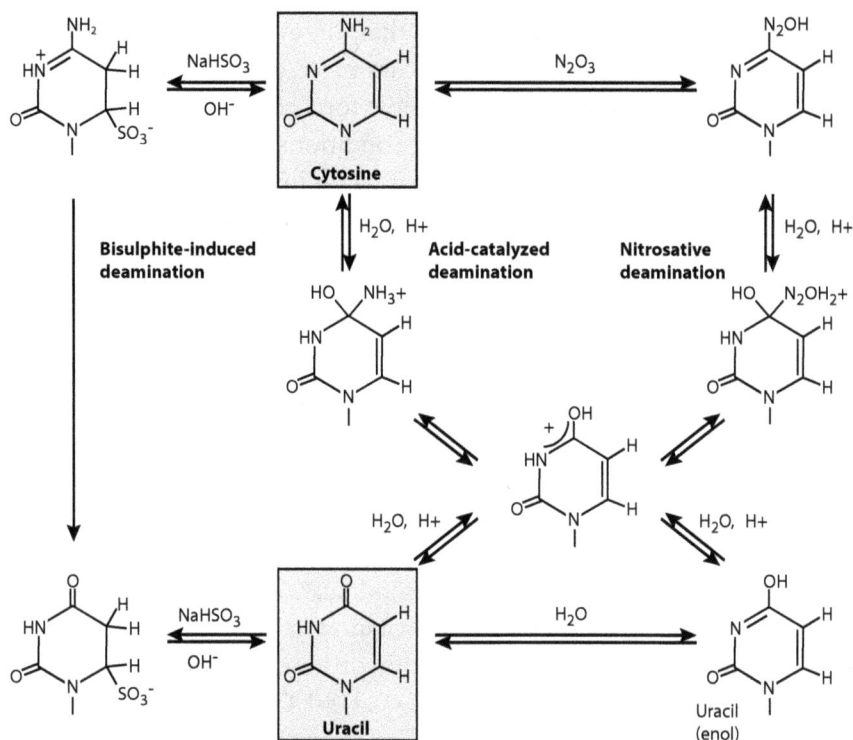

Fig. 3.3. Schemes for spontaneous hydrolytic (middle route), nitrosative (route to the right) and bisulphite-induced deamination of cytosine to uracil (modified from[125]).

of hydrogen peroxide with nitrite, or via reaction of superoxide radical with nitric oxide (NO•). Leakage of superoxide radical from the mitochondrial electron transport chain may thus favor generation of peroxynitrite in mitochondria. The proposed route of nitrosative deamination of cytosine is illustrated in Fig. 3.3.[125] NO• is an important secondary messenger molecule involved in regulating platelet aggregation, blood vessel tone and neurotransmission. It is produced in large amounts during inflammation (up to 10^4 molecules cell^{-1}s^{-1})[126] and may thus constitute a biologically relevant cause of genomic uracil in some cell types. Peroxynitrite-induced deamination occurs 10-fold faster in single-stranded than in double-stranded DNA.[127] Notably, whereas spontaneous deamination of guanine is thought to occur at a much lower rate than that of cytosine, peroxynitrite-induced deamination of guanine occurs twice as fast as that of cytosine,[127] and may therefore be responsible for a considerable fraction of G:C to A:T mutations under certain conditions. Interestingly, peroxynitrite-treated DNA has been shown to be immunogenic, and elicited a prominent antibody response relative to non-treated DNA when injected in rabbits. Moreover, the modified DNA exhibited better epitopes for circulating autoantibodies formed under pathological conditions like cancer and systemic lupus erythematosus (SLE).[128–130] To what degree uracil or intermediates in the cytosine deamination process constitute the neo-epitopes for generation of the autoantibodies remains, however, to be established.

3.2.3. *Deamination of DNA Cytosine by Bisulfite*

In a series of studies on chemical modification of nucleosides and nucleobases in the late 1960s, Hayahatsu *et al.* discovered that cytosine was readily converted to 5,6-dihydrouracil-6-sulfonate by the treatment with $NaHSO_3$ and Na_2SO_3 at 80°C and then further to uracil by treatment with alkali.[131] The proposed reaction steps are outlined in Fig. 3.3. Although not significant under physiological conditions, this reaction is worth mentioning since it has become highly popular to distinguish 5-methylcytosine (5-mC, unaffected by bisulfite) from unmethylated C during DNA[132] and RNA[133] sequencing and has been coined bisulphite sequencing. Since 5-mC is unaffected by bisulfite treatment, only the unmethylated cytosines are converted to uracil, and will give rise to T during sequencing. The positions of 5-mC can then be reconstructed by comparing the bisulfite-treated and non-treated sequencing reactions.

3.2.4. *DNA Cytosine Deamination Induced by UV-irradiation*

Ultraviolet (UV) light may produce genomic uracil, although high doses might be required to reach significant levels compared to other photoproducts. The underlying cause for this is that UV light produces several DNA adducts in which the 5,6-bond of cytosine become saturated and thus makes the base more prone to deamination. This was demonstrated already in 1957, when Green and Cohen found that 5,6-dihydrocytosine deaminated with a half-life of two hours at pH 7.0 and 37°C.[134] The quantitatively dominating monomeric lesion in DNA after UV is 6-hydroxy-5-6-dihydrocytosine (cytosine hydrate),[135] which upon deamination yields uracil hydrate.[136] When poly(dG-[³H]dC) was UV-irradiated, 4% of the cytosine hydrate formed deaminated to uracil hydrate, which was further converted to uracil by slow *trans*-elimination (half-life of 6h at 37°C).[137] A more common source of UV-induced uracil might be deamination of cytosine in cyclopyrimidine dimers (CPDs).[138] Such cytosines deaminate 7–8 orders of magnitude faster than other DNA cytosines and may have a half-life of 5 h or less.[139,140] Conversely, 6,4-photoproducts, which occur at about one third the frequency of CPDs, deaminate nearly two orders of magnitude slower than CPDs.[141] Deamination of cytosine in 5'-CC and 5'-TC CPDs would result in 5'-UU and 5'-TU, respectively, and such deamination has been proposed to be a major contributor to UV-induced mutagenesis.[142] Remarkably, formation of uracils in CPDs was found to follow very unusual kinetics. Whereas no uracils appeared at 37°C within 20 min (single-stranded DNA) and 40 min (double-stranded DNA) following irradiation, most uracils appeared within a brief 14 min period thereafter.[143] To date, this peculiar kinetic pattern remains unexplained.

Direct access of UDGs to deaminated cytosines in CPDs is sterically hindered by the crosslink. In most organisms, this can be circumvented by photoreactivation prior to uracil excision. Photoreactivation occurs by photolyases that harnesses energy from visible light to mediate cleavage of the CPD. This activity appears, however, to be lost in placental mammals, that rather employ nucleotide excision repair to repair CPDs as well as 6,4-photoproducts and a wide range of other "bulky" DNA lesions ([144] and references therein). However, humans have two cryptochromes, CRY1 and CRY2, that share sequence homology with photolyases,[145] and a robust photolyase activity was later demonstrated for other members of the same branch of cryptochromes (Cry-DASH) against CPDs in double-stranded DNA.[146] This raised the possibility that CPD reversal in single-stranded regions such as in front of the replication fork could facilitate uracil removal by replication-associated UNG2. However, no photolyase activity has yet been found in the human CRY proteins, which rather appear to have taken on functions as important regulators of the mammalian circadian clock (reviewed in Refs. 147 and 148). It is also worth mentioning that catalytic DNAs (DNAzymes) have been discovered that were able to use light to mediate photoconversion of thymine dimers.[149,150] The biological relevance of such photoreactivation, if any, remains to be established.

In the 1950s, it was found that 5-bromouracil, which is isosteric with thymine, could be incorporated into DNA when bacteria were incubated with 5-bromo-2'-deoxyuridine (BrdU).[151] It was further found that incorporation of BrdU mediated increased sensitivity to UV and ionizing radiation.[152] This spurred considerable interest into the potential of BrdU and other halogenated pyrimidines as a means to sensitize tumors to radiation therapy.[153] Mechanistic studies suggested that when UV-irradiated, BrdU was converted to the 5-uracil-yl radical as an important intermediate in formation of strand breaks.[154] More recently, BrdU and other halouracils were also shown to induce interstrand[155] as well as intrastrand[156] U-containing crosslinks subsequent to UV-radiation. This has mediated renewed interest in the potential use of halopyrimidines to sensitize tumors harbouring deficiencies in repair of complex DNA damage.

3.2.5. DNA Cytosine Deamination Induced by Ionizing Radiation

Ionizing radiation produces substantial amounts of free-radical products, predominantly reactive oxygen and nitrogen species (ROS and RNS, respectively), including peroxynitrite.[157] The latter has the capability to induce deamination of both cytosine and guanine (see Section 3.2.2). A relationship between ionizing radiation and urinary excretion of 2'-deoxyuridine (dU) in rats was reported in the Russian literature more than 45 years ago[158,159] and in mice subjected to γ radiation, elevated and dose-dependent levels of dU were observed in their urine.[160] To what extent radiation-induced deamination of DNA cytosine contributed to the elevated dU levels, was not addressed. However, in a study from the laboratory of Tomas Lindahl[161] in which [³H]cytosine-labelled DNA was γ-irradiated and treated with either UNG or SMUG1, uracil was identified based on retention time in RP-HPLC. This substantiates that γ-radiation has the capability to induce genomic uracil, most likely by ROS/RNS-mediated deamination, but the levels of genomic uracil generated via this route *in vivo* compared to other mutagenic base lesions, remain to be determined.

3.2.6. *Deamination Induced by Helix-disturbing Agents*

Given that cytosines in single-stranded DNA deaminate much more rapidly than cytosines in double-stranded DNA, agents that induce single-stranded regions in DNA would be expected to mediate increased rate of cytosine deamination in these regions. This has been demonstrated for certain drugs that induce DNA interstrand crosslinks, most notably cisplatin. In the crystal structure of DNA containing a cisplatin interstrand crosslink (ICL) (Fig. 3.4), the DNA is unwound 70–80° and bent towards the minor groove. This causes the cytosines flanking the crosslinked guanines to adopt a extrahelical conformation that exposes them to the nuclear environment.[162] Notably these flipped out cytosines were demonstrated to undergo preferential deamination, creating uracils adjacent to the platinum crosslink.[163] The enzymatic repair of the resulting complex lesion has recently been studied in more detail,[164] and will be discussed in Chapter 4.

3.2.7. *Deamination of Cytosine by Methyltransferases*

About 2–8% of the cytosines in our genome are methylated at the 5-position and this enzymatic DNA methylation occurs

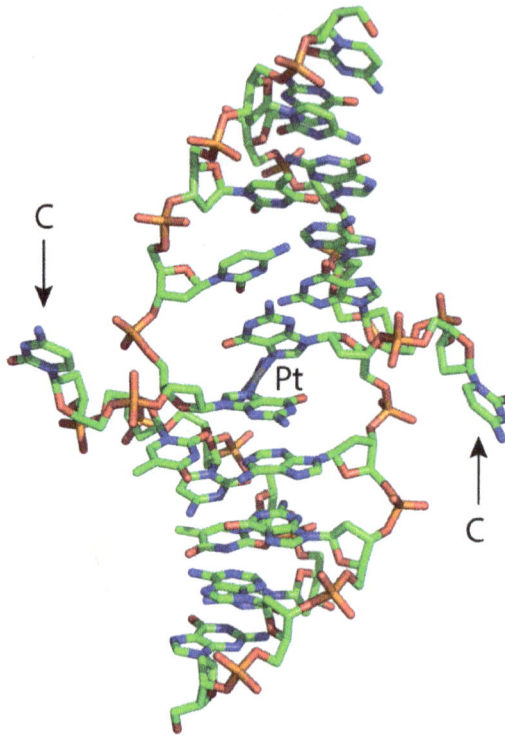

Fig. 3.4. A cisplatin-DNA interstrand crosslink (Pt) mediates extrusion of the flanking cytosines (arrows) from the DNA double helix, rendering them vulnerable to deamination (NDB accession no. DDJ075).

predominantly at CpG sequences. The product, m^5CpG, has an established role in epigenetics by silencing gene expression. Using a plasmid-based *in vitro-in vivo* genetic reversion system, enzymatic deamination of cytosine was first described for the bacterial (cytosine-5) methyltransferase HpaII, which has been reported to induce a 10^4-fold increased rate of reversion (requiring C to T transition) in Ung-deficient bacteria after transformation. Although the mutation frequency was reduced to background levels in the presence of rather low SAM concentrations (300 nM), the occurrence of mutational hot spots at some CpG sites suggests that mutations at such sites might sometimes be caused by enzymes rather than occurring spontaneously.[165] Deamination by DNMTs has largely been studied in bacteria.

However, some eukaryotic DNMTs also deaminate cytosine, at least *in vitro*,[166] and are homologous to their bacterial counterparts.[167] Interestingly, DNMT inhibitors, such as SAM analogs, strongly enhance cytosine deamination,[168-170] as do mutations in the DNMT SAM-binding pocket.[169] However, the possible role of DNMTs in the generation of genomic uracil and associated mutagenesis remains unsubstantiated, but cannot be dismissed, since to our knowledge it has not been sufficiently investigated.

3.2.8. *Deamination of DNA Cytosine by AID/APOBEC Enzymes*

Activation-induced deaminase (AID) and apolipoprotein B mRNA-editing catalytic polypeptide-like (APOBEC) enzymes are present in all tetrapods, including primates and bony fish. They deaminate cytosines either in single-stranded DNA or in both single-stranded DNA and RNA. This source of genomic uracil has in the last decade attracted wide interest because of its important role in adaptive and innate immunity. Moreover, recent high-throughput sequencing of more than 7,000 human cancer genomes identified mutational signatures in many cancer types attributed to the AID/APOBEC family of deaminases,[171] indicating a significant contribution by genomic uracil in mutagenesis in cancer. This important and challenging topic is treated in more detail in Chapters 6 and 7.

References

1. Bessman, M. J., Lehman, I. R., Adler, J., *et al.* Enzymatic synthesis of deoxyribonucleic acid. Iii. The incorporation of pyrimidine and purine analogues into deoxyribonucleic acid. *Proc Natl Acad Sci U S A* **44**, 633–640 (1958).

2. Chargaff, E., Davidson, J. N. *The nucleic acids: chemistry and biology.* 307 (Academic Press, 1955).

3. Zamenhof, S., Reiner, B., De Giovanni, R., Rich, K. Introduction of unnatural pyrimidines into deoxyribonucleic acid of Escherichia coli. *J Biol Chem* **219**, 165–173 (1956).

4. Bertani, L. E., Haggmark, A., Reichard, P. Synthesis of pyrimidine deoxyribonucleoside diphosphates with enzymes from Escherichia coli. *J Biol Chem* **236**, PC67–PC68 (1961).

5. Greenberg, G. R., Somerville, R. L. Deoxyuridylate kinase activity and deoxy-uridinetriphosphatase in Escherichia coli. *Proc Natl Acad Sci U S A* **48**, 247–257 (1962).

6. Takahashi, I., Marmur, J. Replacement of thymidylic acid by deoxyuridylic acid in the deoxyribonucleic acid of a transducing phage for Bacillus subtilis. *Nature* **197**, 794–795 (1963).

7. Kahan, F. M. Novel enzymes formed by Bacillus subtilis infected with bacteriophage. *Federation Proc.* **22**, 406 (1963).

8. Tomita, F., Takahashi, I. A novel enzyme, dCTP deaminase, found in Bacillus subtilis infected with phage PBS I. *Biochim Biophys Acta* **179**, 18–27 (1969).

9. Neuhard, J., Thomassen, E. Deoxycytidine triphosphate deaminase: identification and function in Salmonella typhimurium. *J Bacteriol* **105**, 657–665 (1971).

10. O'Donovan, G. A., Edlin, G., Fuchs, J. A., *et al.* Deoxycytidine triphosphate deaminase: characterization of an Escherichia coli mutant deficient in the enzyme. *J Bacteriol* **105**, 666–672 (1971).

11. Lindahl, T. An N-glycosidase from Escherichia coli that releases free uracil from DNA containing deaminated cytosine residues. *Proc Natl Acad Sci U S A* **71**, 3649–3653 (1974).

12. Konrad, E. B., Lehman, I. R. Novel mutants of Escherichia coli that accumulate very small DNA replicative intermediates. *Proc Natl Acad Sci U S A* **72**, 2150–2154 (1975).

13. Tye, B. K., Nyman, P. O., Lehman, I. R., *et al.* Transient accumulation of Okazaki fragments as a result of uracil incorporation into nascent DNA. *Proc Natl Acad Sci U S A* **74**, 154–157 (1977).

14. Brynolf, K., Eliasson, R., Reichard, P. Formation of Okazaki fragments in polyoma DNA synthesis caused by misincorporation of uracil. *Cell* **13**, 573–580 (1978).

15. Wist, E., Unhjem, O., Krokan, H. Accumulation of small fragments of DNA in isolated HeLa cell nuclei due to transient incorporation of dUMP. *Biochim Biophys Acta* **520**, 253–270 (1978).

16. Grafstrom, R. H., Tseng, B. Y., Goulian, M. The incorporation of uracil into animal cell DNA in vitro. *Cell* **15**, 131–140 (1978).

17. Yoshida, S., Masaki, S. Utilization in vitro of deoxyuridine triphosphate in DNA synthesis by DNA polymerases alpha and beta from calf thymus. *Biochim Biophys Acta* **561**, 396–402 (1979).

18. Dube, D. K., Kunkel, T. A., Seal, G., Loeb, L. A. Distinctive properties of mammalian DNA polymerases. *Biochim Biophys Acta* **561**, 369–382 (1979).

19. Focher, F., Mazzarello, P., Verri, A., *et al.* Activity profiles of enzymes that control the uracil incorporation into DNA during neuronal development. *Mutat Res* **237**, 65–73 (1990).

20. Mosbaugh, D. W. Purification and characterization of porcine liver DNA polymerase gamma: utilization of dUTP and dTTP during in vitro DNA synthesis. *Nucleic Acids Res* **16**, 5645–5659 (1988).

21. Richardson, F. C., Kuchta, R. D., Mazurkiewicz, A., Richardson, K. A. Polymerization of 2'-fluoro- and 2'-O-methyl-dNTPs by human DNA polymerase alpha, polymerase gamma, and primase. *Biochem Pharmacol* **59**, 1045–1052 (2000).

22. Mathews, C. K. DNA precursor metabolism and genomic stability. *FASEB J* **20**, 1300–1314 (2006).

23. Hakansson, P., Hofer, A., Thelander, L. Regulation of mammalian ribonucleotide reduction and dNTP pools after DNA damage and in resting cells. *J Biol Chem* **281**, 7834–7841 (2006).

24. Garavito, M. F., Narvaez-Ortiz, H. Y., Zimmermann, B. H. Pyrimidine Metabolism: Dynamic and Versatile Pathways in Pathogens and Cellular Development. *J Genet Genomics* **42**, 195–205 (2015).

25. Holmgren, A., Reichard, P., Thelander, L. Enzymatic synthesis of deoxyribonucleotides, 8. The effects of ATP and dATP in the CDP reductase system from E. coli. *Proc Natl Acad Sci U S A* **54**, 830–836 (1965).

26. Zhou, B. B., Elledge, S. J. The DNA damage response: putting checkpoints in perspective. *Nature* **408**, 433–439 (2000).

27. Tanaka, H., Arakawa, H., Yamaguchi, T., *et al.* A ribonucleotide reductase gene involved in a p53-dependent cell-cycle checkpoint for DNA damage. *Nature* **404**, 42–49 (2000).

28. Nakano, K., Balint, E., Ashcroft, M., Vousden, K. H. A ribonucleotide reductase gene is a transcriptional target of p53 and p73. *Oncogene* **19**, 4283–4289 (2000).

29. Aye, Y., Li, M., Long, M. J., Weiss, R. S. Ribonucleotide reductase and cancer: biological mechanisms and targeted therapies. *Oncogene* **34**, 2011–2021 (2015).

30. Guittet, O., Hakansson, P., Voevodskaya, N., *et al.* Mammalian p53R2 protein forms an active ribonucleotide reductase in vitro with the R1 protein, which is expressed both in resting cells in response to DNA damage and in proliferating cells. *J Biol Chem* **276**, 40647–40651 (2001).

31. Park, S. J., Kim, H. B., Piao, C., *et al.* p53R2 regulates thioredoxin reductase activity through interaction with TrxR2. *Biochem Biophys Res Commun* **482**, 706–712 (2017).

32. Schaertl, S., Konrad, M., Geeves, M. A. Substrate specificity of human nucleoside-diphosphate kinase revealed by transient kinetic analysis. *J Biol Chem* **273**, 5662–5669 (1998).

33. Bilitou, A., Watson, J., Gartner, A., Ohnuma, S. The NM23 family in development. *Mol Cell Biochem* **329**, 17–33 (2009).

34. Vertessy, B. G., Toth, J. Keeping uracil out of DNA: physiological role, structure and catalytic mechanism of dUTPases. *Acc Chem Res* **42**, 97–106 (2009).

35. Gadsden, M. H., McIntosh, E. M., Game, J. C., *et al.* dUTP pyrophosphatase is an essential enzyme in Saccharomyces cerevisiae. *EMBO J* **12**, 4425–4431 (1993).

36. Tye, B. K., Lehman, I. R. Excision repair of uracil incorporated in DNA as a result of a defect in dUTPase. *J Mol Biol* **117**, 293–306 (1977).

37. Dengg, M., Garcia-Muse, T., Gill, S. G., *et al.* Abrogation of the CLK-2 checkpoint leads to tolerance to base-excision repair intermediates. *EMBO Rep* **7**, 1046–1051 (2006).

38. Kunkel, T. A. Rapid and efficient site-specific mutagenesis without phenotypic selection. *Proc Natl Acad Sci U S A* **82**, 488–492 (1985).

39. Chen, C. W., Tsao, N., Huang, L. Y., *et al.* The Impact of dUTPase on Ribonucleotide Reductase-Induced Genome Instability in Cancer Cells. *Cell Rep* **16**, 1287–1299 (2016).

40. Muha, V., Horvath, A., Bekesi, A., *et al.* Uracil-containing DNA in Drosophila: stability, stage-specific accumulation, and developmental involvement. *PLoS Genet* **8**, e1002738 (2012).

41. Lari, S. U., Chen, C. Y., Vertessy, B. G., *et al.* Quantitative determination of uracil residues in Escherichia coli DNA: Contribution of ung, dug, and dut genes to uracil avoidance. *DNA Repair (Amst)* **5**, 1407–1420 (2006).

42. Hizi, A., Herzig, E. dUTPase: the frequently overlooked enzyme encoded by many retroviruses. *Retrovirology* **12**, 70 (2015).

43. Maley, G. F., Lobo, A. P., Maley, F. Properties of an affinity-column-purified human deoxycytidylate deaminase. *Biochim Biophys Acta* **1162**, 161–170 (1993).

44. Jansen, R. S., Rosing, H., Schellens, J. H., Beijnen, J. H. Deoxyuridine analog nucleotides in deoxycytidine analog treatment: secondary active metabolites? *Fundam Clin Pharmacol* **25**, 172–185 (2011).

45. Marx, A., Alian, A. The first crystal structure of a dTTP-bound deoxycytidylate deaminase validates and details the allosteric-inhibitor binding site. *J Biol Chem* **290**, 682–690 (2015).

46. Munch-Petersen, B., Cloos, L., Tyrsted, G., Eriksson, S. Diverging substrate specificity of pure human thymidine kinases 1 and 2 against antiviral dideoxynucleosides. *J Biol Chem* **266**, 9032–9038 (1991).

47. Valentino, M. L., Marti, R., Tadesse, S., *et al.* Thymidine and deoxyuridine accumulate in tissues of patients with mitochondrial neurogastrointestinal encephalomyopathy (MNGIE). *FEBS Lett* **581**, 3410–3414 (2007).

48. Carreras, C. W., Santi, D. V. The catalytic mechanism and structure of thymidylate synthase. *Annu Rev Biochem* **64**, 721–762 (1995).

49. Chon, J., Stover, P. J., Field, M. S. Targeting nuclear thymidylate biosynthesis. *Mol Aspects Med* **53**, 48–56 (2017).

50. Taddia, L., D'Arca, D., Ferrari, S., *et al.* Inside the biochemical pathways of thymidylate Synthase perturbed by anticancer drugs: Novel strategies to overcome cancer chemoresistance. *Drug Resist Updat* **23**, 20–54 (2015).

51. Lee, L. S., Cheng, Y. Human thymidylate kinase. Purification, characterization, and kinetic behavior of the thymidylate kinase derived from chronic myelocytic leukemia. *J Biol Chem* **252**, 5686–5691 (1977).

52. Su, J. Y., Sclafani, R. A. Molecular cloning and expression of the human deoxythymidylate kinase gene in yeast. *Nucleic Acids Res* **19**, 823–827 (1991).

53. Hu, C. M., Yeh, M. T., Tsao, N., Chen, C. W., Gao, Q. Z., Chang, C. Y., Lee, M. H., Fang, J. M., Sheu, S. Y., Lin, C. J. *et al.* Tumor cells require thymidylate kinase to prevent dUTP incorporation during DNA repair. *Cancer Cell* **22**, 36–50 (2012).

54. Goldstone, D. C., Ennis-Adeniran, V., Hedden, J. J., *et al.* HIV-1 restriction factor SAMHD1 is a deoxynucleoside triphosphate triphosphohydrolase. *Nature* **480**, 379–382 (2011).

55. Li, N., Zhang, W., Cao, X. Identification of human homologue of mouse IFN-gamma induced protein from human dendritic cells. *Immunol Lett* **74**, 221–224 (2000).

56. Lahouassa, H., Daddacha, W., Hofmann, H., *et al.* SAMHD1 restricts the replication of human immunodeficiency virus type 1 by depleting the intracellular pool of deoxynucleoside triphosphates. *Nat Immunol* **13**, 223–228 (2012).

57. Hansen, E. C., Seamon, K. J., Cravens, S. L., Stivers, J. T. GTP activator and dNTP substrates of HIV-1 restriction factor SAMHD1 generate a long-lived activated state. *Proc Natl Acad Sci U S A* **111**, E1843–1851 (2014).

58. Kennedy, E. M., Daddacha, W., Slater, R., *et al.* Abundant non-canonical dUTP found in primary human macrophages drives its frequent incorporation by HIV-1 reverse transcriptase. *J Biol Chem* **286**, 25047–25055 (2011).

59. Rentoft, M., Lindell, K., Tran, P., *et al.* Heterozygous colon cancer-associated mutations of SAMHD1 have functional significance. *Proc Natl Acad Sci U S A* **113**, 4723–4728 (2016).

60. Welbourn, S., Dutta, S. M., Semmes, O. J., Strebel, K. Restriction of virus infection but not catalytic dNTPase activity is regulated by phosphorylation of SAMHD1. *J Virol* **87**, 11516–11524 (2013).

61. Cribier, A., Descours, B., Valadao, A. L., *et al.* Phosphorylation of SAMHD1 by cyclin A2/CDK1 regulates its restriction activity toward HIV-1. *Cell Rep* **3**, 1036–1043 (2013).

62. Pauls, E., Ruiz, A., Badia, R., *et al.* Cell cycle control and HIV-1 susceptibility are linked by CDK6-dependent CDK2 phosphorylation of SAMHD1 in myeloid and lymphoid cells. *J Immunol* **193**, 1988–1997 (2014).

63. Lee, E. J., Seo, J. H., Park, J. H., Vo*et al.* SAMHD1 acetylation enhances its deoxynucleotide triphosphohydrolase activity and promotes cancer cell proliferation. *Oncotarget* **8**, 68517–68529 (2017).

64. Randerath, K., Randerath, E. in *Methods Enzymol* Vol. 12A 323–347 (1967).

65. Flanegan, J. B., Greenberg, G. R. Regulation of deoxyribonucleotide biosynthesis during in vivo bacteriophage T4 DNA replication. Intrinsic control of synthesis of thymine and 5-hydroxymethylcytosine deoxyribonucleotides at precise ratio found in DNA. *J Biol Chem* **252**, 3019–3027 (1977).

66. Hitzeman, R. A., Price, A. R. in *DNA Synthesis: Present and future* (eds I. Molineux, M. Kohiyama) (Plenum Press, 1978).

67. Nilsson, S., Reichard, P., Skoog, L. Deoxyuridine triphosphate pools after polyoma virus infection. *J Biol Chem* **255**, 9552–9555 (1980).

68. Ferraro, P., Franzolin, E., Pontarin, G., *et al.* Quantitation of cellular deoxynucleoside triphosphates. *Nucleic Acids Res* **38**, e85 (2010).

69. Goulian, M., Bleile, B., Tseng, B. Y. The effect of methotrexate on levels of dUTP in animal cells. *J Biol Chem* **255**, 10630–10637 (1980).

70. Jackson, R. C., Jackman, A. L., Calvert, A. H. Biochemical effects of a quinazoline inhibitor of thymidylate synthetase, N-(4-(N-((2-amino-4-hydroxy-6-quinazolinyl)methyl)prop-2-ynylamino) benzoyl)-L-glutamic acid (CB3717), on human lymphoblastoid cells. *Biochem Pharmacol* **32**, 3783–3790 (1983).

71. Piall, E. M., Curtin, N. J., Aherne, G. W., *et al.* The quantitation by radioimmunoassay of 2'-deoxyuridine 5'-triphosphate in extracts of thymidylate synthase-inhibited cells. *Anal Biochem* **177**, 347–352 (1989).

72. Horowitz, R. W., Zhang, H., Schwartz, E. L., *et al.* Measurement of deoxyuridine triphosphate and thymidine triphosphate in the extracts of thymidylate synthase-inhibited cells using a modified DNA polymerase assay. *Biochem Pharmacol* **54**, 635–638 (1997).

73. Studebaker, A. W., Lafuse, W. P., Kloesel, R., Williams, M. V. Modulation of human dUTPase using small interfering RNA. *Biochem Biophys Res Commun* **327**, 306–310 (2005).

74. Webley, S. D., Hardcastle, A., Ladner, R. D., *et al.* Deoxyuridine triphosphatase (dUTPase) expression and sensitivity to the thymidylate synthase (TS) inhibitor ZD9331. *Br J Cancer* **83**, 792–799 (2000).

75. Maybaum, J., Klein, F. K., Sadee, W. Determination of pyrimidine ribotide and deoxyribotide pools in cultured cells and mouse liver by high-performance liquid chromatography. *J Chromatogr* **188**, 149–158 (1980).
76. James, S. J., Miller, B. J., Basnakian, A. G., *et al.* Apoptosis and proliferation under conditions of deoxynucleotide pool imbalance in liver of folate/methyl deficient rats. *Carcinogenesis* **18**, 287–293 (1997).
77. Halsted, C. H., Villanueva, J., Chandler, C. J., *et al.* Ethanol feeding of micropigs alters methionine metabolism and increases hepatocellular apoptosis and proliferation. *Hepatology* **23**, 497–505 (1996).
78. Lawrance, A. K., Deng, L., Rozen, R. Methylenetetrahydrofolate reductase deficiency and low dietary folate reduce tumorigenesis in Apc min/+ mice. *Gut* **58**, 805–811 (2009).
79. Fromentin, E., Gavegnano, C., Obikhod, A., Schinazi, R. F. Simultaneous quantification of intracellular natural and antiretroviral nucleosides and nucleotides by liquid chromatography-tandem mass spectrometry. *Anal Chem* **82**, 1982–1989 (2010).
80. Chen, X., McAllister, K. J., Klein, B., *et al.* Development and validation of an LC-MS/MS quantitative method for endogenous deoxynucleoside triphosphates in cellular lysate. *Biomed Chromatogr* (2016).
81. Otterlei, M., Warbrick, E., Nagelhus, T. A., *et al.* Post-replicative base excision repair in replication foci. *EMBO J* **18**, 3834–3844 (1999).
82. Nilsen, H., Rosewell, I., Robins, P., *et al.* Uracil-DNA glycosylase (UNG)-deficient mice reveal a primary role of the enzyme during DNA replication. *Mol Cell* **5**, 1059–1065 (2000).
83. Galashevskaya, A., Sarno, A., Vagbo, C. B., *et al.* A robust, sensitive assay for genomic uracil determination by LC/MS/MS reveals lower levels than previously reported. *DNA Repair (Amst)* **12**, 699–706 (2013).
84. Rona, G., Scheer, I., Nagy, K., *et al.* Detection of uracil within DNA using a sensitive labeling method for in vitro and cellular applications. *Nucleic Acids Res* (2015).
85. Shin, Y. S., Chan, C., Vidal, A. J., Brody, *et al.* Subcellular localization of gamma-glutamyl carboxypeptidase and of folates. *Biochim Biophys Acta* **444**, 794–801 (1976).
86. Prem veer Reddy, G., Pardee, A. B. Multienzyme complex for metabolic channeling in mammalian DNA replication. *Proc Natl Acad Sci U S A* **77**, 3312–3316 (1980).
87. veer Reddy, G. P., Pardee, A. B. Coupled ribonucleoside diphosphate reduction, channeling, and incorporation into DNA of mammalian cells. *J Biol Chem* **257**, 12526–12531 (1982).

88. veer Reddy, G. P., Pardee, A. B. Inhibitor evidence for allosteric interaction in the replitase multienzyme complex. *Nature* **304**, 86–88 (1983).

89. Reddy, G. P. Compartmentation of deoxypyrimidine nucleotides for nuclear DNA replication in S phase mammalian cells. *J Mol Recognit* **2**, 75–83 (1989).

90. Noguchi, H., Prem veer Reddy, G., Pardee, A. B. Rapid incorporation of label from ribonucleoside disphosphates into DNA by a cell-free high molecular weight fraction from animal cell nuclei. *Cell* **32**, 443–451 (1983).

91. Wickremasinghe, R. G., Yaxley, J. C., Hoffbrand, A. V. Gel filtration of a complex of DNA polymerase and DNA precursor-synthesizing enzymes from a human lymphoblastoid cell line. *Biochim Biophys Acta* **740**, 243–248 (1983).

92. Anderson, D. D., Woeller, C. F., Chiang, E. P., *et al.* Serine hydroxymethyltransferase anchors de novo thymidylate synthesis pathway to nuclear lamina for DNA synthesis. *J Biol Chem* **287**, 7051–7062 (2012).

93. Wan, C., Borgeson, B., Phanse, S., *et al.* Panorama of ancient metazoan macromolecular complexes. *Nature* **525**, 339–344 (2015).

94. Niida, H., Katsuno, Y., Sengoku, M., *et al.* Essential role of Tip60-dependent recruitment of ribonucleotide reductase at DNA damage sites in DNA repair during G1 phase. *Genes Dev* **24**, 333–338 (2010).

95. Jarrett, S. G., Novak, M., Dabernat, S., *et al.* Metastasis suppressor NM23-H1 promotes repair of UV-induced DNA damage and suppresses UV-induced melanomagenesis. *Cancer Res* **72**, 133–143 (2012).

96. Woeller, C. F., Anderson, D. D., Szebenyi, D. M., Stover, P. J. Evidence for small ubiquitin-like modifier-dependent nuclear import of the thymidylate biosynthesis pathway. *J Biol Chem* **282**, 17623–17631 (2007).

97. Sarangi, P., Zhao, X. SUMO-mediated regulation of DNA damage repair and responses. *Trends Biochem Sci* **40**, 233–242 (2015).

98. Tang, Y., Schon, E. A., Wilichowski, E., *et al.* Rearrangements of human mitochondrial DNA (mtDNA): new insights into the regulation of mtDNA copy number and gene expression. *Mol Biol Cell* **11**, 1471–1485 (2000).

99. Rampazzo, C., Ferraro, P., Pontarin, G., *et al.* Mitochondrial deoxyribonucleotides, pool sizes, synthesis, and regulation. *J Biol Chem* **279**, 17019–17026 (2004).

100. Di Noia, M. A., Todisco, S., Cirigliano, A., *et al.* The human SLC25A33 and SLC25A36 genes of solute carrier family 25 encode two mitochondrial pyrimidine nucleotide transporters. *J Biol Chem* **289**, 33137–33148 (2014).

101. Wang, L. Mitochondrial purine and pyrimidine metabolism and beyond. *Nucleosides Nucleotides Nucleic Acids* **35**, 578–594 (2016).

102. Lynx, M. D., Bentley, A. T., McKee, E. E. 3'-Azido-3'-deoxythymidine (AZT) inhibits thymidine phosphorylation in isolated rat liver mitochondria: a possible mechanism of AZT hepatotoxicity. *Biochem Pharmacol* **71**, 1342–1348 (2006).

103. Jullig, M., Eriksson, S. Mitochondrial and submitochondrial localization of human deoxyguanosine kinase. *Eur J Biochem* **267**, 5466–5472 (2000).

104. Van Rompay, A. R., Johansson, M., Karlsson, A. Phosphorylation of nucleosides and nucleoside analogs by mammalian nucleoside monophosphate kinases. *Pharmacol Ther* **87**, 189–198 (2000).

105. McKee, E. E., Bentley, A. T., Hatch, M., *et al.* Phosphorylation of thymidine and AZT in heart mitochondria: elucidation of a novel mechanism of AZT cardiotoxicity. *Cardiovasc Toxicol* **4**, 155–167 (2004).

106. Ferraro, P., Nicolosi, L., Bernardi, P., *et al.* Mitochondrial deoxynucleotide pool sizes in mouse liver and evidence for a transport mechanism for thymidine monophosphate. *Proc Natl Acad Sci U S A* **103**, 18586–18591 (2006).

107. Chen, Y. L., Lin, D. W., Chang, Z. F. Identification of a putative human mitochondrial thymidine monophosphate kinase associated with monocytic/macrophage terminal differentiation. *Genes Cells* **13**, 679–689 (2008).

108. Milon, L., Meyer, P., Chiadmi, M., *et al.* The human nm23-H4 gene product is a mitochondrial nucleoside diphosphate kinase. *J Biol Chem* **275**, 14264–14272 (2000).

109. Kamath, V. G., Hsiung, C. H., Lizenby, Z. J., McKee, E. E. Heart mitochondrial TTP synthesis and the compartmentalization of TMP. *J Biol Chem* **290**, 2034–2041 (2015).

110. Anderson, D. D., Quintero, C. M., Stover, P. J. Identification of a de novo thymidylate biosynthesis pathway in mammalian mitochondria. *Proc Natl Acad Sci U S A* **108**, 15163–15168 (2011).

111. Bogenhagen, D. F., Rousseau, D., Burke, S. The layered structure of human mitochondrial DNA nucleoids. *J Biol Chem* **283**, 3665–3675 (2008).

112. Ladner, R. D., Caradonna, S. J. The human dUTPase gene encodes both nuclear and mitochondrial isoforms. Differential expression of the isoforms and characterization of a cDNA encoding the mitochondrial species. *J Biol Chem* **272**, 19072–19080 (1997).

113. Yan, N., O'Day, E., Wheeler, L. A., *et al.* HIV DNA is heavily uracilated, which protects it from autointegration. *Proc Natl Acad Sci U S A* **108**, 9244–9249 (2011).

114. Weil, A. F., Ghosh, D., Zhou, Y., *et al.* Uracil DNA glycosylase initiates degradation of HIV-1 cDNA containing misincorporated dUTP and prevents viral integration. *Proc Natl Acad Sci U S A* **110**, E448–457 (2013).

115. Schuster, H., Schramm, G. [Method of chemical determination of the biologically active unit of ribonucleic acid in tobacco mosaic virus]. *Z Naturforsch B* **13B**, 697–704 (1958).

116. Hurst, R. O., Kuksis, A. Degradation of deoxyribonucleic acid by hot alkali. *Can J Biochem Physiol* **36**, 919–929 (1958).

117. Shapiro, R., Klein, R. S. The deamination of cytidine and cytosine by acidic buffer solutions. Mutagenic implications. *Biochemistry* **5**, 2358–2362 (1966).

118. Freese, E. B. Transitions and transversions induced by depurinating agents. *Proc Natl Acad Sci U S A* **47**, 540–545 (1961).

119. Lindahl, T. Irreversible heat inactivation of transfer ribonucleic acids. *J Biol Chem* **242**, 1970–1973 (1967).

120. Lindahl, T., Nyberg, B. Heat-induced deamination of cytosine residues in deoxyribonucleic acid. *Biochemistry* **13**, 3405–3410 (1974).

121. Shapiro, R. in *Chromosome Damage and Repair, Ed's Erling Seeberg and Kjell Kleppe* Vol. 40 3–18 (Plenum Press, ISBN 0-306-40886-4, New York London, 1980).

122. Kavli, B., Otterlei, M., Slupphaug, G., Krokan, H. E. Uracil in DNA--general mutagen, but normal intermediate in acquired immunity. *DNA Repair (Amst)* **6**, 505–516 (2007).

123. Merchant, K., Chen, H., Gonzalez, T. C., *et al.* Deamination of single-stranded DNA cytosine residues in aerobic nitric oxide solution at micromolar total NO exposures. *Chem Res Toxicol* **9**, 891–896 (1996).

124. Wink, D. A., Kasprzak, K. S., Maragos, C. M., *et al.* DNA deaminating ability and genotoxicity of nitric oxide and its progenitors. *Science* **254**, 1001–1003 (1991).

125. Labet, V., Grand, A., Morell, C., *et al.* Mechanism of nitric oxide induced deamination of cytosine. *Phys Chem Chem Phys* **11**, 2379–2386 (2009).

126. Tamir, S., Burney, S., Tannenbaum, S. R. DNA damage by nitric oxide. *Chem Res Toxicol* **9**, 821–827 (1996).

127. Caulfield, J. L., Wishnok, J. S., Tannenbaum, S. R. Nitric oxide-induced deamination of cytosine and guanine in deoxynucleosides and oligonucleotides. *J Biol Chem* **273**, 12689–12695 (1998).

128. Habib, S., Moinuddin, Ali, R. Acquired antigenicity of DNA after modification with peroxynitrite. *Int J Biol Macromol* **35**, 221–225 (2005).

129. Habib, S., Moinuddin, Ali, R. Peroxynitrite-modified DNA: a better antigen for systemic lupus erythematosus anti-DNA autoantibodies. *Biotechnol Appl Biochem* **43**, 65–70 (2006).

130. Habib, S., Moinuddin, Ali, A., Ali, R. Preferential recognition of peroxynitrite modified human DNA by circulating autoantibodies in cancer patients. *Cell Immunol* **254**, 117–123 (2009).

131. Hayatsu, H., Wataya, Y., Kazushige, K. The addition of sodium bisulfite to uracil and to cytosine. *J Am Chem Soc* **92**, 724–726 (1970).

132. Frommer, M., McDonald, L. E., Millar, D. S., *et al.* A genomic sequencing protocol that yields a positive display of 5-methylcytosine residues in individual DNA strands. *Proc Natl Acad Sci U S A* **89**, 1827–1831 (1992).

133. Schaefer, M., Pollex, T., Hanna, K., Lyko, F. RNA cytosine methylation analysis by bisulfite sequencing. *Nucleic Acids Res* **37**, e12 (2009).

134. Green, M., Cohen, S. S. Studies on the biosynthesis of bacterial and viral pyrimidines. II. Derivatives of dihydrocytosine. *J Biol Chem* **228**, 601–609 (1957).

135. Liu, F. T., Yang, N. C. Photochemistry of cytosine derivatives. 2. Photohydration of cytosine derivatives. Proton magnetic resonance study on the chemical structure and property of photohydrates. *Biochemistry* **17**, 4877–4885 (1978).

136. Johns, H., LeBlanc, J., Freeman, K. Reversal and deamination rates of the main ultraviolet photoproduct of cytidylic acid. *J Mol Biol* **13**, 849–861 (1965).

137. Boorstein, R. J., Hilbert, T. P., Cunningham, R. P., Teebor, G. W. Formation and stability of repairable pyrimidine photohydrates in DNA. *Biochemistry* **29**, 10455–10460 (1990).

138. Fix, D., Bockrath, R. Thermal resistance to photoreactivation of specific mutations potentiated in E. coli B/r ung by ultraviolet light. *Mol Gen Genet* **182**, 7–11 (1981).

139. Barak, Y., Cohen-Fix, O., Livneh, Z. Deamination of cytosine-containing pyrimidine photodimers in UV-irradiated DNA. Significance for UV light mutagenesis. *J Biol Chem* **270**, 24174–24179 (1995).

140. Setlow, R. B., Carrier, W. L., Bollum, F. J. Pyrimidine dimers in UV-irradiated poly dI:dC. *Proc Natl Acad Sci U S A* **53**, 1111–1118 (1965).

141. Douki, T., Cadet, J. Far-UV photochemistry and photosensitization of 2'-deoxycytidylyl-(3'-5')-thymidine: isolation and characterization of the main photoproducts. *J Photochem Photobiol B* **15**, 199–213 (1992).

142. Tessman, I., Liu, S. K., Kennedy, M. A. Mechanism of SOS mutagenesis of UV-irradiated DNA: mostly error-free processing of deaminated cytosine. *Proc Natl Acad Sci U S A* **89**, 1159–1163 (1992).

143. Tessman, I., Kennedy, M. A., Liu, S. K. Unusual kinetics of uracil formation in single and double-stranded DNA by deamination of cytosine in cyclobutane pyrimidine dimers. *J Mol Biol* **235**, 807–812 (1994).

144. Sugasawa, K. Molecular mechanisms of DNA damage recognition for mammalian nucleotide excision repair. *DNA Repair (Amst)* **44**, 110–117 (2016).

145. Brudler, R., Hitomi, K., Daiyasu, H., *et al.* Identification of a new crypto-chrome class. Structure, function, and evolution. *Mol Cell* **11**, 59–67 (2003).

146. Selby, C. P., Sancar, A. A cryptochrome/photolyase class of enzymes with single-stranded DNA-specific photolyase activity. *Proc Natl Acad Sci U S A* **103**, 17696–17700 (2006).

147. Gustafson, C. L., Partch, C. L. Emerging models for the molecular basis of mammalian circadian timing. *Biochemistry* **54**, 134–149 (2015).

148. Kavakli, I. H., Baris, I., Tardu, M., *et al.* The Photolyase/Cryptochrome Family of Proteins as DNA Repair Enzymes and Transcriptional Repressors. *Photochem Photobiol* **93**, 93–103 (2017).

149. Chinnapen, D. J., Sen, D. A deoxyribozyme that harnesses light to repair thymine dimers in DNA. *Proc Natl Acad Sci U S A* **101**, 65–69 (2004).

150. Thorne, R. E., Chinnapen, D. J., Sekhon, G. S., Sen, D. A deoxyribozyme, Sero1C, uses light and serotonin to repair diverse pyrimidine dimers in DNA. *J Mol Biol* **388**, 21–29 (2009).

151. Zamenhof, S., Gribiff, G. E. coli containing 5-bromouracil in its deoxyribo-nucleic acid. *Nature* **174**, 307–308 (1954).

152. Djordjevic, B., Szybalski, W. Genetics of human cell lines. III. Incorporation of 5-bromo- and 5-iododeoxyuridine into the deoxyribonucleic acid of human cells and its effect on radiation sensitivity. *J Exp Med* **112**, 509–531 (1960).

153. Kinsella, T. J., Mitchell, J. B., Russo, A., *et al.* The use of halogenated thymi-dine analogs as clinical radiosensitizers: rationale, current status, and future prospects: non-hypoxic cell sensitizers. *Int J Radiat Oncol Biol Phys* **10**, 1399–1406 (1984).

154. Sugiyama, H., Tsutsumi, Y., Saito, I. Highly sequence-selective photoreaction of 5-bromouracil-containing deoxyhexanucleotides. *J Am Chem Soc* **112**, 6720–6721 (1990).

155. Cecchini, S., Girouard, S., Huels, M. A., *et al.* Interstrand cross-links: a new type of gamma-ray damage in bromodeoxyuridine-substituted DNA. *Biochemistry* **44**, 1932–1940 (2005).

156. Zeng, Y., Wang, Y. Sequence-dependent formation of intrastrand crosslink products from the UVB irradiation of duplex DNA containing a 5-bromo-2'-deoxyuridine or 5-bromo-2'-deoxycytidine. *Nucleic Acids Res* **34**, 6521–6529 (2006).

157. Saenko, Y., Cieslar-Pobuda, A., Skonieczna, M., Rzeszowska-Wolny, J. Changes of reactive oxygen and nitrogen species and mitochondrial functioning in human K562 and HL60 cells exposed to ionizing radiation. *Radiat Res* **180**, 360–366 (2013).

158. Mazurik, V. K., Bryksina, L. E., Bibikhin, L. N. [The relationship between the excretion of deoxyuridine, thymidine and beta-aminoisobutyric acid by rats and the radiation dose and length of time following total irradiation]. *Radiobiologiia* **10**, 43–48 (1970).

159. Mazurik, V. K., Bryksina, L. E., Saprygin, D. B., Iarilin, A. A. [A study of the role of lymphoid tissue in post-radiation hyperexcretion of deoxycytidine, deoxyuridine and thymidine by use of specific antisera]. *Radiobiologiia* **10**, 346–349 (1970).

160. Tyburski, J. B., Patterson, A. D., Krausz, K. W., *et al.* R. Radiation metabolomics. 2. Dose- and time-dependent urinary excretion of deaminated purines and pyrimidines after sublethal gamma-radiation exposure in mice. *Radiat Res* **172**, 42–57 (2009).

161. An, Q., Robins, P., Lindahl, T., Barnes, D. E. C → T mutagenesis and gamma-radiation sensitivity due to deficiency in the Smug1 and Ung DNA glycosylases. *EMBO J* **24**, 2205–2213 (2005).

162. Coste, F., Malinge, J. M., Serre, L., *et al.* Crystal structure of a double-stranded DNA containing a cisplatin interstrand cross-link at 1.63 A resolution: hydration at the platinated site. *Nucleic Acids Res* **27**, 1837–1846 (1999).

163. Kothandapani, A., Dangeti, V. S., Brown, A. R., *et al.* Novel role of base excision repair in mediating cisplatin cytotoxicity. *J Biol Chem* **286**, 14564–14574 (2011).

164. Sawant, A., Floyd, A. M., Dangeti, M., *et al.* Differential role of base excision repair proteins in mediating cisplatin cytotoxicity. *DNA Repair (Amst)* (2017).

165. Shen, J. C., Rideout, W. M., 3rd, Jones, P. A. High frequency mutagenesis by a DNA methyltransferase. *Cell* **71**, 1073–1080 (1992).

166. Metivier, R., Gallais, R., Tiffoche, C., *et al.* Cyclical DNA methylation of a transcriptionally active promoter. *Nature* **452**, 45–50 (2008).

167. Kumar, S., Cheng, X., Klimasauskas, S., *et al.* The DNA (cytosine-5) methyltransferases. *Nucleic Acids Res* **22**, 1–10 (1994).

168. Sharath, A. N., Weinhold, E., Bhagwat, A. S. Reviving a dead enzyme: cytosine deaminations promoted by an inactive DNA methyltransferase and an S-adenosylmethionine analogue. *Biochemistry* **39**, 14611–14616 (2000).

169. Stier, I., Kiss, A. Cytosine-to-uracil deamination by SssI DNA methyltransferase. *PLoS One* **8**, e79003 (2013).

170. Zingg, J. M., Shen, J. C., Yang, A. S., *et al.* Methylation inhibitors can increase the rate of cytosine deamination by (cytosine-5)-DNA methyltransferase. *Nucleic Acids Res* **24**, 3267–3275 (1996).

171. Alexandrov, L. B., Nik-Zainal, S., Wedge, D. C., *et al.* Signatures of mutational processes in human cancer. *Nature* **500**, 415–421 (2013).

Chapter 4

Enzymology of Genomic Uracil Repair

Hans E. Krokan*, Bodil Kavli, Antonio
Sarno and Geir Slupphaug

Whereas several laboratories contributed to the discovery of other DNA excision repair processes, the discovery of base excision repair (BER) can be uniquely ascribed to the work of Tomas Lindahl (Fig. 4.1). He quantified spontaneous deamination of cytosine to uracil in genomic DNA and reasoned that there had to be a mecshanism correcting the highly mutagenic U:G mismatches resulting from the deamination.[1] The search for such a mechanism resulted in the surprising discovery of uracil-DNA glycosylase in 1974,[2] the first enzyme in a new family of DNA repair proteins. The new enzyme was initially named 'uracil N-glycosidase,' but the name was later changed to uracil-DNA glycosylase to comply with current nomenclature. In addition to his work on DNA-cytosine deamination, Lindahl made important discoveries on other aspects of DNA instability, including rates of depurination, depyrimidination and chain breaks at abasic sites.[3-6]

Prior to his work on DNA stability, Tomas Lindahl worked on tRNA in the laboratory of Jacques R. Fresco at Princeton University. There, Lindahl contributed significant papers on the thermal denaturation, the dynamic structure and stability of tRNA, as well as the

---

* hans.krokan@ntnu.no

Fig. 4.1. Professor Tomas Lindahl at the Royal Swedish Academy of Sciences. He was rewarded The Nobel Prize in Chemistry in 2015 together with Paul Modrich and Aziz Sancar "for mechanistic studies of DNA repair." (*Source*: Holger Motzkau — Creative Commons.)

significance of Mg^{2+} in these processes. Thus, he had made himself familiar with the stability and structural properties of RNA, as well as the significance of the physical and chemical environment of RNA, prior to his work on the stability of DNA. Important discoveries in science are usually inspired by earlier discoveries and this was probably also so in the case of cytosine deamination. As early as in 1958, Hurst and Kuksis demonstrated that in hot alkali (1 N NaOH, 100°C, 4 hrs) cytosine in DNA was fully degraded, 50% of which found as the deamination product uracil.[7] Furthermore, the work of Robert Shapiro and his group had previously demonstrated deamination of the isolated base cytosine and the nucleoside cytidine under rather mild acidic or alkaline conditions.[8]

For almost two decades, the type of uracil-DNA glycosylase discovered by Tomas Lindahl[2] and later cloned from herpesvirus,[9] *E. coli*[10] and human cells,[11] was the only known type of UDG. Then uracil-DNA glycosylases TDG and SMUG1 were independently identified. Genes for human UNG, TDG and SMUG1 are all located on the long arm of chromosome 12, and as elucidated in Chapter 2, they appear to have evolved from a common ancestor. In 1993 and 1996 the laboratory of Josef Jiricny demonstrated that a mammalian T:G mismatch-binding protein identified in 1987[12] was in fact a DNA glycosylase that removed thymine as well as uracil in T:G and U:G mismatches. This protein was distinctly different from the previously identified uracil-DNA glycosylases[13,14] and was later named TDG, a Family 2 type UDG. It is present in both eukaryotes and prokaryotes. Some years later uracil-DNA glycosylase SMUG1 (Single-strand Selective Monofunctional Uracil-DNA Glycosylase) was identified in *Xenopus laevis* (Haushalter et al., 1999). This type of protein, constituting the Family 3 type of UDG, was thought to be vertebrate-specific, but is now known to be present in both eukaryote and prokaryote cells and it is usually not single-strand specific. As the number of sequenced organisms has grown, new genes encoding verified or putative UDGs have been identified. Family 4 UDG was first identified and characterized biochemically in the thermophilic bacterium *Thermotoga maritima* but analogues were identified in several other organisms, including some human-pathogens.[15] Family 5 UDG was identified as a broad specificity UDG in the hyperthermophilic archaen *Pyrobaculum aerophilum*.[16] The list of verified and putative UDGs continues to grow and is interesting both in an evolutionary and a functional perspective. The present text will, however, focus on genomic uracil and its processing in mammalian cells and a biomedical context. Furthermore, we will start the overview by elucidating some chronological aspects of research in the field, as the UDGs were not assigned to specific families at that time.

4.1. DNA Glycosylases are Ubiquitous

In the discussion of his 1974-paper on uracil DNA glycosylase Lindahl states: "One possible function of this enzyme, acting in concert with endonuclease II, an exonuclease, a DNA polymerase, and DNA ligase, would be the reversion of guanine-uracil base pairs in DNA to guanine-cytosine pairs by excision-repair." Thus, the very first paper on DNA glycosylases in principle outlined the complete pathway for repair of genomic uracil. The outline of this pathway according to our present view is illustrated in Fig. 4.2.

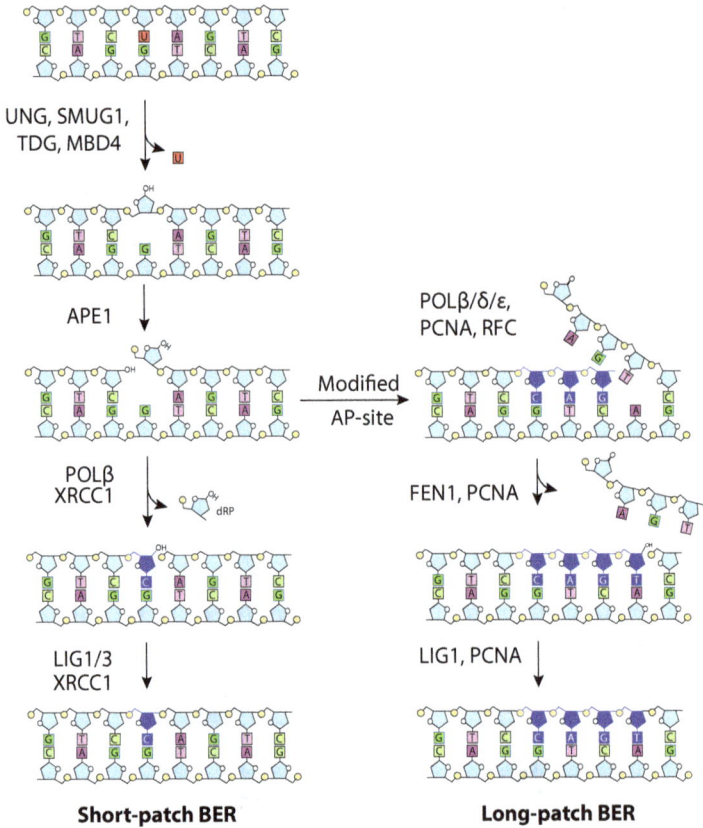

Fig. 4.2. Outline of the base-excision repair pathway for genomic uracil according to our current understanding. The core enzymatic steps of this pathway were suggested by Tomas Lindahl in 1974. The long-patch pathway uses several DNA replication proteins and is largely active in proliferating cells.

The second enzyme in the proposed pathway, endonuclease II from *E. coli*, was isolated as early as 1969 and found to degrade alkylated DNA,[17] but was later found to be an AP-endonuclease.[18] The reason why degraded alkylated DNA was a good substrate *in vitro* compared with native DNA is probably that alkylated purines 7-methylguanine and 3-methyladenine in DNA are rapidly lost by spontaneous hydrolytic depurination. Importantly, Lindahl had also been central in the discovery or characterization of other enzymes in the novel pathway, including AP-endonucleases and ligases.[19] Later his laboratory discovered DNA glycosylases recognizing other base lesions. These include a DNA glycosylase that could remove 3-methyladenine,[20] one that could remove hypoxanthine (deamination product of adenine)[21] and one that could remove ring-opened 7-methylguanine residues from DNA, chemically named 2,6-diamino-4-hydroxy-5-N-methylformamido-pyrimidine ("FaPy"). FaPy-DNA glycosylase had properties principally similar to other DNA glycosylases that Lindahl had discovered by being relatively small, globular and with no requirement for divalent ions.[22] These findings established DNA glycosylases as a class of enzymes initiating base excision repair of a variety of base lesions, so uracil-DNA glycosylase was not just a rare exception. Presently, 11 different mammalian DNA glycosylases are known, some of which having nuclear and mitochondrial isoforms. Jointly, these DNA-glycosylases remove a wide spectrum of deaminated, oxidized, alkylated, partially degraded or mismatched bases.[23]

Uracil-DNA glycosylase in *E. coli* also removes incorporated uracil-residues in U:A pairs resulting from incorporation of dUTP in cells.[24-26] In isolated nuclei from human cells, uracil-DNA glycosylase rapidly removes incorporated uracil from nuclear DNA in a post-replicative process resulting in generation of newly synthesized DNA fragments smaller than normal Okazaki fragments,[27,28] although normal-size Okazaki fragments do not appear to result from uracil-removal.[28] Subsequent work has identified four mammalian DNA glycosylases that have the capacity to remove uracil, as well as certain other damaged pyrimidines from DNA. These are generally known as UNG1 (mitochondrial) and UNG2 (nuclear), SMUG1, TDG and MBD4 and have overlapping as well as distinct properties.[23,29]

Since DNA glycosylases, unlike nucleases, do not require a divalent cation for activity, DNA-glycosylase activity may be measured under conditions where nucleases are inactive. Monofunctional DNA glycosylases simply remove the damaged base, leaving an abasic site.[2] Bifunctional glycosylases also have a lyase activity that may cleave DNA by β-elimination at the abasic site.[23] DNA glycosylases removing uracil, alkylation lesions or a mismatched base are monofunctional, whereas those removing oxidative damage are usually bifunctional.

Following the discovery of uracil-DNA glycosylase in *E. coli*, similar activities were detected in *B. subtilis*,[30] *Micrococcus luteus*,[31] herpesvirus,[32] poxvirus,[33] the yeast *Saccharomyces cerevisae*,[34] the brine shrimp *Artemia salina*,[35] cod,[36] wheat germ,[37] carrots,[38] mouse,[39] human cells[28,40,41] and numerous other species.[42,43]

4.2. Bacterial and Yeast Uracil-DNA Glycosylases

The work on purified bacterial and yeast uracil-DNA glycosylases established them as relatively small, monomeric and Mg^{2+}-independent enzymes that release uracil from DNA without cleaving the DNA backbone. The Lindahl laboratory purified *E. coli* uracil-DNA glycosylase (Ung) 11,000-fold from 80 g of cells harvested during logarithmic growth. The enzyme preparation contained ~ 99% uracil-DNA glycosylase protein and was thus close to homogeneity.[24] *E. coli* Ung is a monomeric protein of molecular weight 24.5 ± 1 kDa. The enzyme release uracil bonded to either adenine (U:A pairs) or guanine (U:G mismatches) in DNA, does not require co-factors and is independent of Mg^{2+} or other divalent cations. It is also active on single stranded DNA, which may be considered unexpected, since base excision repair requires a complementary strand for the repair process. It does not act on DNA containing 5-bromouracil, pyrimidine dimers or deaminated purine residues; neither are RNA or free dUMP substrates. Ung introduces apyrimidinic sites in covalently closed circular DNA molecules containing a few deaminated cytosine residues, but does not cleave the phosphodiester bond itself, as demonstrated using neutral and alkaline sucrose gradients. Such sucrose gradients are informative, since "relaxed" circular molecules (with a single-strand break) and closed circles sediment differently in such gradients and because apyrimidinic sites are alkali-sensitive. A partially purified Ung protein from *B. subtilis* had very similar properties.[30] Furthermore, an essentially homogenous preparation of uracil-DNA glycosylase from the yeast *Saccharomyces cerevisiae* had properties very similar to the corresponding enzyme from *E. coli*, although the yeast enzyme is slightly larger (27 kDa).[34]

The type of mutation expected to increase in cells unable to correct U:G mismatches is C → T transitions. To examine the role of Ung, *E. coli* cells deficient in Ung-activity were isolated using a non-selective mass screening procedure combined with activity measurements. The *ung* gene was mapped between *tyrA* and *nadB* on the *E. coli* chromosome. Mutants were viable and grew normally, thus Ung is per definition nonessential. The most deficient mutant (less than 0.02% remaining Ung-activity) displayed a 4.5-5.4-fold increased mutation frequency, indicating a relatively modest role of Ung in mutation avoidance.[44] However, later *trpA* reversion studies revealed a 30-fold increased rate of G:C → A:T transitions.[45] Similarly, a different laboratory found approximately 10-fold increased mutation frequency in *E. coli* cells deficient in Ung.[46] A possible explanation for this relatively modest increase in mutation frequency could be a uracil-processing backup activity. A mismatch specific uracil-DNA glycosylase identified more than a decade later (Mug in *E. coli*, TDG in human cells) was expected to be such a backup. This type of enzyme removes T or U mismatched to G and was first identified and cloned from human cells.[13,14] A homologous enzyme was subsequently identified in *E. coli* and some other species.[47] Although Mug would appear to be a very likely backup based on its biochemical properties, Mug surprisingly did not seem to have a significant role in uracil-removal in *E. coli*, but appeared to be important in removal of 3,N^4-ethenocytosine (εC).[46] εC is a DNA-adduct generated by lipid peroxidation products and the industrial toxicant vinyl chloride.[48] However, whereas expression of Ung is high in exponentially growing cells, it is very low in stationary-phase cultures. In contrast, the expression of Mug is low in proliferating cells and high in stationary-phase cells and it was demonstrated to have an important role in mutation avoidance in stationary cells.[49] Thus, Mug and Ung appear to complement each other in different growth phases of *E. coli*.

4.3. Purification and cDNA Cloning of Eukaryote Uracil-DNA Glycosylases

Following the discovery of Ung in *E. coli* many efforts were made to purify a corresponding mammalian enzyme, for a number of years unsuccessfully. This was largely due to low abundance in tissues and cells examined, and an apparent heterogeneity during chromatography. Initial work demonstrated that the properties of human UNG were largely similar to those of *E. coli* Ung in partially purified preparations; 10-fold from human fibroblasts[41] and 50-80-fold from HeLa cells.[28,50] Elution profiles during chromatography indicated heterogeneity, in retrospect most likely caused by molecular interactions and posttranslational modifications. Similar properties were also observed with partially purified preparations of human nuclear UNG (70% of total activity), mitochondrial UNG (15% of total activity) purified approximately 920-fold and 250-fold, respectively, from HeLa cells. Some 15% of total UNG activity was found in the cytosol.[51] These results confirmed earlier reports on UNG activity in nuclear and mitochondrial extracts of human cells.[52,53] Purification of UNG using blast cells from patients with acute myelocytic leukemia resulted in a 1,000-fold purification of an enzyme that was reportedly essentially homogenous, with an apparent molecular weight of 30,000 and a specific activity of 688 units/mg protein (1 unit releases 1 nmol uracil/min at 37°C).[54] While it cannot be excluded that UNG has a different activity in this cell type, or that a fraction of UNG was inactive, the reported specific activity is very low compared to that of an apparently homogenous UNG protein purified from HeLa cells later.[55] Using a large amount of HeLa cells as starting material (3.6×10^{10} cells), the latter study took advantage of a more extensive purification procedure to obtain a monomeric uracil-DNA glycosylase of molecular weight 29,000 and a specific activity of 17,313 units/mg protein (1 unit releases 1 nmol uracil/min at 30°C). This result indicated that human uracil-DNA glycosylase is a relatively low abundance protein, even in proliferating cells, and that purification to homogeneity of protein amounts that would allow further characterization required rather extensive measures in terms of amount of material and number of purification steps required.

Early attempts at cloning cDNAs for human uracil-DNA glycosylase employed antibodies raised against UNG preparations that may not have been homogenous. These efforts resulted in identification of cDNAs encoding a presumed uracil-DNA glycosylase identical to a subunit of glyceraldehyde 3-phosphate dehydrogenase (GAPDH) of molecular weight 37,000 Daltons[56,57] and a cyclin-like uracil-DNA glycosylase.[58,59] While it cannot be excluded that GAPDH and the cyclin-like protein could carry cryptic or minor uracil-DNA glycosylase activity, a rather overwhelming majority of scientific papers demonstrate that uracil-DNA glycosylase activity in mammalian cells is encoded by quite different genes, as described in subsequent sections in this chapter and subsequent chapters. Importantly, rather standard biochemical methods were used to detect uracil-DNA glycosylase activity in the papers that led to the identification of GAPDH and a cyclin-like protein as uracil-DNA glycosylases. Yet, these papers failed to identify any of the genes that are now known to encode *bona fide* uracil-DNA glycosylases.

4.4. Uracil-DNA Glycosylase Encoded by *UNG* — Family 1 UDGs

Uracil-DNA glycosylases encoded by UNG-genes constitute the Family 1 UDGs, present in all domains of life.[42] Since human UNG is actually a relatively low abundance protein, purification of uracil-DNA glycosylase required large amounts starting material, with placenta being a good candidate. After optimizing purification procedures in pilot experiments, the purification process was scaled up by starting with 20 kg of human placentae. Using nine chromatography steps, 131,000-fold purification was achieved starting from a crude extract that had been centrifuged and filtered through a DEAE-cellulose column. After the final FPLC Mono S step, 60 μg of protein with very high uracil-DNA glycosylase activity (8,473 units/mg) was obtained and found to contain two protein bands by SDS-polyacrylamide slab gel electrophoresis. Both bands were blotted onto Polybrene glass fibre sheets and subjected to amino acid analyses as well as amino acid sequencing. A sequence of 26 amino acids near the N-terminal end of the band presumed to contain uracil-DNA glycosylase was obtained.[60] The sequences of both proteins represented proteins that at that time were different from known proteins. The protein that co-purified with uracil-DNA glycosylase has later been identified as a polypyrimidine tract binding protein.

The N-terminal amino acid sequence of the presumed uracil-DNA glycosylase was used to design degenerate probes for screening of a cDNA library. This resulted in cloning of a cDNA encoding uracil-DNA glycosylase.[11] The protein encoded by this cDNA complemented *E. coli ung* mutants.[61] The encoded protein had striking homology to the corresponding uracil-DNA glycosylase genes recently cloned from herpes simplex virus 1 and 2,[9] an open reading frame of Varicella-Zoster (gene 59),[62] a poxvirus,[33,63] *E. coli*[10] and yeasts.[64] These results demonstrated clearly that the *E. coli ung* gene is one representative of a large family (UNG) of viral, prokaryotic and eukaryotic genes encoding uracil-DNA glycosylase activities. In terms of enzymatic activity, it is the major uracil-DNA glycosylase both in prokaryotes and eukaryotes and the only known uracil-DNA glycosylase encoded by virus. Furthermore, the rather extensive conservation explains the earlier observation that the bacteriophage PBS2-encoded peptide Ugi, which is a very strong inhibitor of *B. subtilis* uracil-DNA glycosylase, also efficiently inhibits uracil-DNA glycosylase from *E. coli*, yeast, human nuclei and mitochondria, but does not inhibit other DNA glycosylases.[65] It was also an early indication that nuclear and mitochondrial uracil-DNA glycosylase were related.

4.4.1. *The Human UNG Gene Encodes Nuclear and Mitochondrial Forms*

The relatively small *UNG*-gene (~13.8 kb) was cloned from a bacteriophage P1-library and found to be located in position 12q24.1.[66] Two different forms of cDNA encoding mouse UNG were subsequently identified and shown to encode mitochondrial (UNG1) and nuclear (UNG2) forms of UNG differing at the N-terminal end. Sequencing of the corresponding mouse gene identified an exon (1a) that encoded the unique N-terminal end of UNG2. Reanalysis of the human gene sequence identified a nearly identical exon (1a) in humans.[67] UNG1 and UNG2 thus have

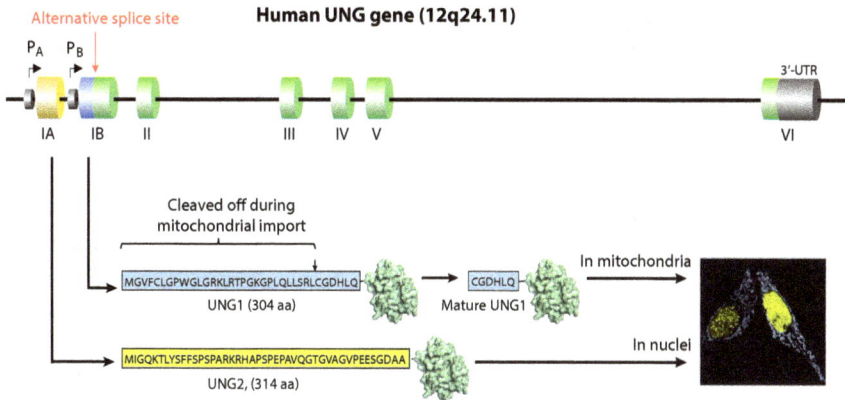

Fig. 4.3. The UNG gene located in chromosome position 12q24.11 spans about 13.5 Kb and contains seven exons. Promoters P_A and P_B is used to generate transcripts for nuclear (UNG2) and mitochondrial (UNG1) isoforms, respectively. The nuclear transcript is generated by splicing of exon 1A (yellow) into the alternative splice site of exon 1B, thus skipping the upstream (blue) part of this exon. All of exon 1B is used to generate UNG1, resulting in a unique N-terminal sequence. During translocation of UNG1 into mitochondria, the 29 amino acid N-terminal mitochondrial targeting sequence is cleaved off by the mitochondrial processing peptidase, resulting in a mature UNG1 protein of 275 amino acids.[70]

unique N-terminal sequences, but identical catalytic domains and the enzymatic properties are very similar (Fig. 4.3). Unlike other DNA glycosylases, including uracil-DNA glycosylases SMUG1, TDG and MBD4,[29] UNG1 and 2 have highest activity with single-stranded DNA substrate (Table 4.1), but are also very active with double-stranded DNA[50,68] and has turnover numbers at least 1–3 orders of magnitude higher than SMUG1, TDG and MBD4, depending on assay conditions.[29,68,69] Importantly, UNG2 is the only nuclear DNA glycosylase that efficiently removes uracil from U:A pairs close to the advancing replication fork,[71,72] a function that this enzyme is ideally suited for due to its high turnover number. In agreement with this function, it is largely located in replication foci during S-phase, but virtually excluded from nucleoli.[68,72,73] The catalytic properties of UNG also suggest that it is the major activity removing uracil from U:G mismatches.[68,69] However, this has not been directly documented except in the case of U:G

Table 4.1. Mammalian Uracil DNA Glycosylases

Enzyme	Subcellular Localization	Substrate Bases/Context[1]	Cell Cycle Regulation[2]	Mouse Knockout	Human Disease
UNG1	Mitochondria	Same as UNG2		Unknown	Unknown
UNG2[3]	Nuclei Replication foci	Uss>U:G>U:A (5-FU, 5-hU, 5-caU, alloxan, isodialuric acid)	Peaks in S phase	Partial defect in CSR and SHM, B-cell lymphomas	HIGM syndrome Defective CSR Lymphoid hyperplasia
SMUG1	Nuclei Nucleoli Cajal bodies (Cytoplasm)	U:G>U:A>Uss 5-hmU, 5-hU, 5-fU (alloxan, isodialuric acid, 5-FU, εC)	No cell cycle regulation	Viable and fertile, increased levels of 5-hmU in brain	Unknown
TDG	Nuclei	U, T, 5-hmU, Tg, 5-fC, 5-caC, εC, 5-FU opposite G	Peaks in G1, not in S phase	Embryonic lethal, epigenetic role in development	Unknown
MBD4 (MED1)	Nuclei	U, T, 5-hmU, Tg εC, 5-FU opposite G in CpG context	No cell cycle regulation	Viable and fertile, increased C to T transitions at CpG intestinal neoplasia	Mutated in carcinomas with MSI

[1]*Minor substrates are in brackets;* [2](Ref. 121), [3]*Mouse knockout (Ung*[−/−]*) and human disease (mutations affecting UNG1 and UNG2); 5-FU (5-fluorouracil); 5-hU (5-hydroxyuracil); 5-hmU (5-hydroxymethyluracil); 5-caU (5-carboxyuracil); 5-fC (5-formylcytosine); 5-caC (5-carboxycytosine); εC (ethenocytosine); Tg (Thymine glycol); HIGM (hyper-IgM syndrome); MSI (microsatellite instability), see text for references.*

mismatches generated by activation-induced cytosine deaminase (AID). Such U:G mismatches are key intermediates in adaptive immunity. This is treated in more detail in Chapter 6 and recapitulated briefly here. During the immune response, activated B-cells generate antibody diversity by somatic hypermutation (SHM) and class switch recombination (CSR). The molecular mechanisms of SHM and CSR remained a mystery until AID was discovered in 1999–2000,[74,75] although initially assumed to be an RNA-cytosine deaminase. AID deaminates cytosine in DNA in specific regions of the *Ig* loci. These DNA-uracils are substrates for UNG2, as demonstrated by the molecular characteristics of UNG deficient mice and human patients. During CSR, UNG2 together with AP endonuclease is necessary to convert clustered AID-generated uracils in the switch regions to double-strand breaks. This triggers recombination by a non-homologous end joining process and connects Ig variable region with a new constant region encoding IgG, IgA or IgE (Chapter 6, Fig. 6.7). Moreover, UNG2-generated abasic sites in IgG variable regions are substrates for translesion synthesis (TLS) by error prone polymerases during SHM[76] (Chapter 6, Fig. 6.6).

UNG1/2 also has the capacity to remove some oxidation products of cytosine from DNA, such as alloxan, isodialuric acid and 5-hydroxyuracil,[77] but it remains unknown whether this has significance in *in vivo* processing of such oxidation products. In addition, UNG2 possesses weak 5-carboxyuracil excision activity *in vitro*.[78] Interestingly, recent evidence indicates that UNG2 may be involved in TET-mediated demethylation of cytosine. Co-expression of UNG2 together with TET2 reactivated expression from a methylation-silenced reporter plasmid and reduced the level of genomic 5-carboxycytosine in transfected HEK293T cells. Moreover, a partial deficit in demethylation at specific genomic loci in of *Ung*[−/−] mouse zygotes was observed, supporting a non-canonical role of UNG2 in epigenetic regulation.[78]

UNG1 contains a mitochondrial targeting sequence that is removed during import to mitochondria,[70] leaving an N-terminally processed protein containing the catalytic domain. Apparently, UNG1 does not engage in multiprotein complexes.[79] The N-terminal extension of UNG2 is part of a complex nuclear localization signal[80] and engages in direct interactions with RPA and PCNA, while XRCC1 appears to bind outside of the N-terminal region.[72,73,81–84] In addition, UNG2-complexes contain additional repair and replication factors, including POLβ, POLδ, APE1 and a DNA ligase activity.[81] The N-terminal extension of UNG2 is also subject to at least three phosphorylations during progression of the S-phase. The initial phosphorylation of S23 at the G1/S phase boundary increases association with chromatin and catalytic turnover, whereas later phosphorylations at T60 and S64 reduce this affinity and may prepare UNG2 for breakdown in late S/G2 phase.[73]

4.4.2. *Structure of the UNG Protein and Mechanism of Catalysis*

Crystallization of the compact catalytic domain of human UNG in combination with extensive site-directed mutagenesis revealed the overall structure, the DNA-binding groove, a compact uracil-binding pocket and active site residues critical for catalytic activity, and suggested a catalytic mechanism. Briefly, the topology of UNG is that of a classic α/β protein containing a central four-stranded sheet of parallel β strands surrounded by eight α helices. A positively charged groove of width equal to that of double-stranded DNA represents the DNA-binding groove. The active site uracil-binding pocket is located towards the C-terminal edge of the parallel β-sheet and can only accommodate uracil if it is flipped out of the helix.[85,86] The herpes simplex 1 (HSV1) UNG protein is structurally closely related.[87]

Importantly, these studies were also the first to report base-flipping (actually, nucleotide flipping, see below) in DNA repair (Fig. 4.4A). Uracil is flipped into a tight catalytic pocket and held in place by main chain and side chain interactions with uracil, base stacking and shape complementarity. Residues critical for catalytic activity cluster in the uracil-binding pocket. Using UNG1-numbering of amino acid residues, main chain N of Gln144 and Asp145 form H-bonds with O2 in uracil, whereas the side chain amide of Asn204 interacts with N3 and O4. Tyr147 forms a lid preventing access of thymine due to the size of 5-methyl in thymine, whereas Phe158 stacks with uracil (Fig. 4.4B). The nucleophilic attack by the imidazole of His268 on the C1 bond is facilitated by these extensive interactions. Single mutation of each of these residues causes loss of more than 99% (but not 100%) of the

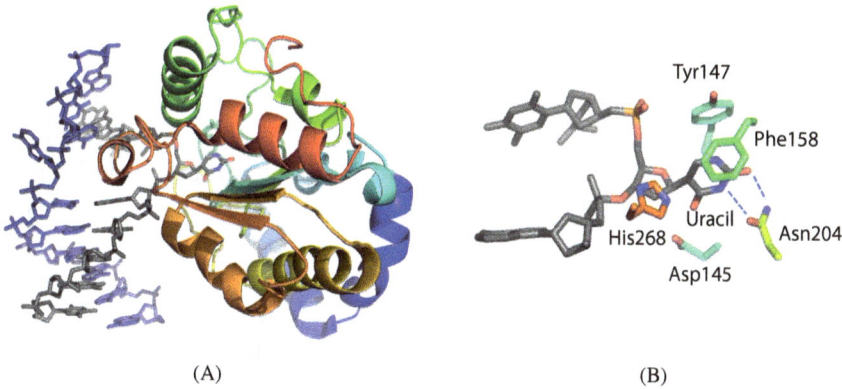

Fig. 4.4. (A) Prior to excision of uracil by UNG, the dUMP moiety is flipped about 180° from the DNA helix and accommodated into the narrow uracil-binding pocket of the glycosylase. (B) Active site residues in human UNG1/2 that contribute to catalysis and substrate specificity (PDB ID: 1SSP). See text for details.

enzyme activity. The discrimination against removal of thymine by Tyr147 was verified through substituting this residue by an amino acid carrying a smaller side chain (e.g. Ala). This mutant released undamaged thymine from both single stranded and double stranded DNA. Similarly, the Asn204Asp mutant lost discrimination against cytosine removal.[88] In cells, these mutants generate a natural abasic site opposite either adenine or guanine and have been used to study the mutagenicity and repair of abasic sites in bacteria[89] and yeast.[90] Moreover, transgenic mouse models with inducible and cell-specific expression of the UNG1 Tyr147Ala mutant were used to study mitochondrial abasic site-triggered neuronal damage[91,92] and cardiac dysfunction.[93]

The mechanism described above focuses on active site chemistry in the uracil-binding pocket and does not take into consideration the effects of the much more extensive interactions between UNG and target DNA.[86,94] At least four motifs in UNG are involved in the steps leading to substrate binding and uracil removal. These are the Leu272 loop (268-HPSPLSVYR-276) involved in flipping and/or stabilization of the enzyme-product complex, the 4-Pro loop (165-PPPPS-169) and the Gly-Ser loop (246-GS-247) anchoring 5′- and 3′-phosphates flanking deoxyuridine and contributing to "pinching", the uracil specificity β2 region (201-LLLN-204) involved in uracil-binding, and the water-activating loop (145-DPYH-148) directly required for catalysis in the uracil-binding pocket and exclusion of thymine.[94] In addition, many other residues interact with substrate DNA more loosely and contribute to the initial contact. One limitation in the first published structures was that the N-C1' glycosylic bond between uracil and deoxyribose was always cleaved, even with catalytically impaired mutants of UNG. Importantly, the deoxyribose and flanking phosphates remained in the flipped out positions both with cleaved U-DNA and abasic site DNA as substrate.[86,94] What was missing was the structure of an enzyme-substrate complex with the glycosylic bond intact. The solution was high-resolution co-crystal structures of UNG and DNA with a non-hydrolysable C-C glycosylic bond in deoxypseudouridine (dψU) (1.8 Å). This structure was compared with that of UNG and control with a slowly cleavable analog with an N-C1' glycosylic bond.[95] These structures strongly indicate that the extensive interactions between enzyme and substrate prior to cleavage lead to a conformational strain on the glycosylic bond that weakens it and contributes to catalysis. The enzyme structure resembles that of the "closed" product complex, with Leu272 inserted in the base stack without cleavage of the glycosylic bond. In the flipped out state, the uracil-ring is rotated approximately 90° on its N1-C4 axis and uracil is raised to a stretched out semi-axial position, moving it halfway from the normal *anti*-conformation towards a *syn*-position. This alters the substrate stereochemistry. Prior to substrate-enzyme binding the glycosylic N1-C1' σ-bond and π-orbital in uracil C2-oxygen were orthogonal to one another (not interacting). However, in the strained conformation these orbitals align, facilitating electron transposition and glycosylic bond cleavage. These findings also suggest that activated water may possibly be less central in catalysis than originally assumed and that conformational strain in the extensive interaction between DNA-substrate and UNG is important. Possibly, the significance of this type of substrate-enzyme strain may be limited to interactions between very large substrate molecules and their enzymes.

4.4.3. *The Mechanism of Base Flipping*

Flipping of the target base and the catalytic mechanism of DNA glycosylases have been the topic of several subsequent papers and reviews,[86–88,94–99] but the mechanism is still not fully understood. Base flipping as a biochemical mechanism was first demonstrated for HhaI methyltransferase that methylates cytosine to 5-methylcytosine in DNA.[100] In base excision repair, it was found that uracil had to be flipped out from the DNA helix in order to be accommodated in the tight uracil-binding pocket.[85,87] However, these papers did not directly demonstrate the enzyme in complex with uracil flipped out from a double stranded DNA molecule. Direct demonstration of base flipping was first achieved for human UNG in complex with damaged target DNA.[86,94,95] Deoxyribose was also flipped out along with flanking phosphates, thus nucleotide flipping is a more appropriate term than base flipping.[86] The positively charged DNA binding groove of width 21 Å (similar to double-stranded DNA) makes numerous electrostatic contacts with the DNA substrate. In the crystal structure, the side chain of Arg in the catalytically impaired Leu272Arg mutant is inserted into DNA through the minor groove and uracil flipped out via the major groove. As shown later, the side chain of Leu in Leu272 is similarly inserted,[94] even in a DNA with an uncleavable C-C glycosylic bond.[95] Furthermore, the DNA backbone is compressed and bent through interactions between several residues in the enzyme and backbone phosphates and deoxyribose. The crystal structure and binding kinetics for wild type and mutants (Leu272Arg, Leu272Ala and Leu272Arg/Asp145Asn) suggested that the side chain of Leu272 is important for uracil-DNA binding and catalysis, in spite of being positioned at a distance >10 Å from the uracil-binding pocket. Uracil is bound in the active site pocket through multiple main chain and side chain interactions as described in Section 4.4.2. The flipped out uracil moves ~15 Å from the intra-helical position to the flipped out position with uracil remaining in the active site even after the base-sugar bond is broken. These studies suggested a model involving enzyme-assisted flipping ("push and pull") via the major groove by complex interactions between UNG and DNA.[86] However, subsequent studies on the catalytically impaired Leu272Ala mutant demonstrated that uracil was flipped out also in the Leu272Ala-target DNA complex, although the Ala side chain is too short to fill the hole in the base stack. Furthermore, deoxyribose was flipped out even in the complex of Leu272Ala and abasic site-DNA instead of U-DNA. Thus, the UNG mutant Leu272Ala can detect and flip abasic sites, demonstrating that minor groove and backbone interactions are sufficient for nucleotide flipping and that neither the pull on uracil from the binding pocket nor push from Leu272 is essential for nucleotide flipping.[94] However, these findings do not rule out a possible contribution of push and pull in the very rapid catalysis by wild-type enzyme and normal substrate. In the updated model on the initial UNG-target DNA contact, UNG binds to DNA and kinks and compresses the backbone slightly with a 'Ser-Pro pinch' and scans the minor groove for damage. Upon identification of a uracil-containing nucleotide, it is flipped out and into the UNG active site, which results in more pronounced DNA binding, slight contraction of the active site pocket and efficient cleavage of the N-C1′ glycosylic bond, leaving UNG bound to the abasic site-DNA.[94] Continued binding until AP-endonuclease cleaves the abasic site DNA is advantages due to its toxicity.

However, the complex enzyme substrate interactions described above do not easily explain how UNG searches for the rare DNA-uracils in a vast excess of undamaged bases. Short-range sliding of the enzyme (perhaps in a "search conformation") along DNA may take place without extensive interactions and may rapidly interrogate bases that are temporarily extrahelical. If the base is undamaged, the base is rapidly released. If damaged, a more specific excision complex is formed and the base excised,[97] somewhat reminiscent of the model above. The relative excess of enzyme over uracil-DNA may allow three-dimensional scanning by enzyme-jumping from one strand to another one in the vicinity. It is not a requirement for the glycosylase function that the base pair is unstable or in a mismatch, because certain tailored active site mutants have the capacity to remove perfectly normal pyrimidines from double-stranded (and single-stranded DNA).

Thus, the Asn204Asp mutant can release cytosine from C:G base pairs, whereas the Tyr147Ala mutant removes thymine from T:A pairs.[88] However, it remains to be determined unequivocally whether UNG can transiently open perfect base pairs or whether it depends on spontaneous and very short-lived openings of base pairs.

4.4.4. *Characteristics of UNG-deficiency*

Ung[-/-] mice are fully viable, fertile, have apparently normal phenotype, grow normally and have only slightly increased overall mutation rates, as measured by the Big Blue system.[71,101] However, they have compromised class switch recombination (CSR) and distorted somatic hypermutation,[102] increased lymphoid hyperplasia, often monoclonal, and a ~20-fold increased frequency of B-cell lymphomas.[103,104] In addition, MEFs have increased genomic uracil.[101,105,106] In humans UNG-deficiency causes lymphoid hyperplasia, recurrent infections and much more severe CSR-deficiency than in mice.[107] This species difference may be explained by the large difference in relative abundance of UNG and SMUG1 in humans compared with mice.[69] Furthermore, using a middle cerebral artery occlusion-reperfusion model, Ung-deficient mice were found to develop larger brain infarctions compared with wild-type and were more sensitive to neuronal cell death.[101] Folate deficiency increases mutation frequencies in *Ung*[-/-] MEFs and results in increased death of cultured hippocampal cells, as well as increased CA3 pyramidal cell death in the hippocampus of mice.[108] One year survival is not reduced in UNG or SMUG1 knockout mice. Somewhat surprisingly, it is also not reduced in UNG-SMUG1 double knockouts. In contrast, MSH2 knockout very significantly reduces survival. The survival is further slightly/moderately reduced in MSH2-UNG double knockouts and very severely reduced in MSH2-UNG-SMUG1 triple knockouts, the major cause of death being lymphomas.[109] These results suggest that genomic uracil is counteracted by several defense mechanisms, although the major role of MSH2 is most likely in repair of mismatches other than U:G. In conclusion, UNG-deficiency causes a severely altered phenotype, particularly when analysed under cellular stress, with changes being most apparent in the immune system and the brain.

4.4.5. *Additional Roles of UNG*

In addition to its major role in genomic uracil sanitation, UDGs of the UNG family apparently have important functions in several other contexts. The role of mammalian UNG in mediating SHM and CSR was briefly described above and will be covered in more detail in Chapter 6. The same chapter will also outline the contribution of mammalian UNG to innate immunity by its apparent degradation of proviral DNA that has been uracilated by APOBEC type cytosine deaminases. Some viruses, like herpes- and poxviruses, encode their own UNG proteins that are required for virulence, reactivation from latency and even constitute part of viral replication machineries. Other viruses, like HIV-1, do not encode their own UNG, but may hijack cellular UNG into virions. The potential functions of UNG proteins in these contexts will be described in more detail in Chapter 5. Finally, UNG from *Xenopus laevis* and humans has been found to contribute to assembly of the histone H3 variant and kinetochore-assembly protein CENP-A.[110,111] Whereas inhibition of UNG2 in Xenopus egg extracts was sufficient to block assembly of CENP-A in sperm nuclei, addition of UNG2 mutants with promiscuous substrate specificities induced additional CENP-A foci. This indicates that the abasic sites generated by UNG2, rather than uracil itself, underlie CENP-A-recruitment. Addition of dUTPase to the egg extracts did not affect CENP-A loading, suggesting that centromeric uracil origi-nating from misincorporation of dUMP was not the underlying cause of CENP-A-recruitment. Conversely, zebularine, a cytidine deami-nase inhibitor, completely blocked CENP-A foci formation.[111] Although zebularine has other functions in addition to deaminase inhibition, this may suggest the intriguing possibility that CENP-A loading and thus kinetochore assembly might involve targeted induc-tion of centromeric uracil by a yet unidentified cytidine deaminase. Thus, further investigation of the molecular mechanisms involved in centromere uracil formation and its functional consequences are clearly warranted.

4.5. T/U:G-Mismatch Glycosylase — The Family 2 UDGs

Family 2 UDGs remove mismatched uracils (U:G) and thymines (T:G) from DNA and are frequently called TDG (in mammalian cells) and Mug (in prokaryotes). There is evidence implicating dysregulated or mutant TDGs in cancer, but this is not necessarily related to the role of the enzyme in BER, as discussed in Chapter 7, Sections 7.1 and 7.5. Similarly to UNG and SMUG1, the gene for TDG is located on the long arm of chromosome 12 in human cells (position 12q23.3). Importantly, TDG/Mug family proteins also have roles in transcription and demethylation and deletion or inactivation of TDG is embryonic lethal. TDG was identified as a protein that bound to T:G mismatches that may arise from 5-meC deamination and subsequently purified, cloned at the cDNA level and demonstrated to be a thymine-mismatch DNA glycosylase.[12–14] Furthermore, bacterial and mammalian TDGs are more efficient with U:G mismatches as substrate as compared with T:G,[47] although TDG in general exhibit a very slow turnover rate. TDG also removes εC caused by lipid peroxidation and 5-hmU paired with G (Table 4.1). TDG binds very tightly to the AP:G mismatch resulting from the glycosylase action and TDG is SUMOylated to help dissociation from the abasic site.[112] The order of declining affinity for DNA is AP:G>U:G>T:G>C:G.[113] Interestingly TDG is highly expressed during G2-M and G1 phases of the cell cycle, but not during the S-phase when the UNG2 protein level is high.[114] The presence of TDG during S-phase may be disadvantageous to the cell because slow turnover and strong binding to abasic sites may stall the replication machinery.

TDG knockout in mice is embryonically lethal and associated with aberrant promoter methylation and imbalanced histone modifications, impairing the expression of developmental genes.[115] TDG associates with promoters, interacts with several transcriptional factors, transcriptional coactivators and DNA methyltransferases, supporting a role of TDG in transcriptional regulation.[116] Moreover, TDG is directly coupled to active DNA demethylation in mammalian cells by excising 5-formylC (5-fC) and 5-carboxyC (5-caC) generated by TET-mediated oxidation of 5-mC, followed by BER.[117,118]

4.6. Family 3 UDG — SMUG1 Removes Hydroxymethyluracil (5-hmU) in DNA and Complements UNG in Uracil Removal

SMUG1 was originally identified in *X. laevis* and coined Single-strand-selective Monofunctional Uracil-DNA Glycosylase (xSMUG1) because it preferred single-stranded DNA as substrate.[119] However, in the presence of salts, human and mouse SMUG1 are essentially double strand-specific.[69] Like genes for UNG2 and TDG, the gene for SMUG1 is located in the long arm of chromosome 12, at position 12q13.13. In spite of less than 10% amino acid similarity, UNG, TDG and SMUG1 have similar structures and may have evolved from the same ancestral gene.[120] In contrast to the *UNG* gene, *SMUG1* is not cell cycle regulated.[121] SMUG1 was thought to solely be a vertebrate DNA glycosylase, but is actually also found in several bacteria.[42,122]

Unlike UNG2, SMUG1 does not accumulate in replication foci, but is rather evenly distributed in nuclei and accumulates in nucleoli and its expression level is independent of the cell cycle stage. In addition, it is also present in cytosol.[68] The order of preference for uracil-substrates for human and mouse SMUG1 is U:G>U:A>U in ssDNA (Table 4.1).[68,69,123] At least *in vitro*, SMUG1 is a catalytically slow enzyme, particularly compared with UNG.[68] In mice, SMUG1 was reported to be the major enzyme to initiate BER of U:G mismatches, while the role of UNG was suggested to be limited to U:A repair.[124] Moreover, mouse SMUG1 has been assigned backup roles for UNG in both SHM and CSR (see Chapter 6). However, the level of expression of SMUG1 relative to UNG is much higher in mouse than in man,[69] and there is to our knowledge no evidence supporting a unique role of SMUG1 in processing U:G mismatches in human cells. SMUG1 has been found to interact with pseudouridine synthase dyskerin (DKC1) with which it co-localizes in nucleoli and Cajal bodies. In addition, SMUG1 was found to have activity on 5-hmU in single-stranded RNA, while depletion of SMUG1 reduced the level of mature rRNA. These observations suggest a role of SMUG1 in quality control of RNA.[125]

4.6.1. *Smug1 is Essential for 5-hmU-repair and Complements Ung in Uracil Removal*

Somewhat surprisingly, SMUG1 and SMUG1-UNG double knockouts are fertile and have the same one-year survival as wild-type mice. However, survival is dramatically reduced in SMUG1-UNG-MSH2 triple knockouts compared with MSH2 knockout alone, suggesting that mismatch repair may function as a last resort to remove genomic uracil.[109] The content of genomic uracil in mouse organs in *Smug1* single knockouts was found to be indistinguishable from that of wild-type mice, whereas UNG single knockouts had 1.9-2.2-fold increases. However, in SMUG1-UNG double knockouts a dramatic increase in genomic uracil was observed, particularly in liver (25-fold increase).[126] These results strongly indicate that SMUG1 and UNG complement each other, although SMUG1 cannot fully complement UNG-deficiency. However, SMUG1 is the major DNA glycosylase removing 5-hmU from DNA.[68,109,127,128] Expectedly, SMUG1 knockouts accumulated 5-hmU in genomic DNA. However, the accumulation of 5-hmU was several-fold higher in the brain than in other tissues. UNG knockouts had normal 5-hmU excision activity and did not accumulate genomic 5-hmU. Furthermore, SMUG1 single knockouts and SMUG1-UNG double knockouts displayed similar 5-hmU excision activity and genomic 5-hmU accumulation.[126] These results, together with other results discussed in this chapter and Chapter 6, demonstrate that UNG and SMUG1 have both unique and overlapping functions. As examples, UNG has unique functions in adaptive immunity, whereas SMUG1 has a unique function in removing 5-hmU from DNA, and even from single stranded RNA. Table 4.1 summarizes results related to DNA repair by UNG and SMUG1.

4.7. MBD4 — A DNA Glycosylase with a Methyl-CpG Binding Domain

MBD4, also known as MED1, is unique among the uracil-DNA glycosylases due to its N-terminal methyl-binding domain (MBD) connected to the C-terminal glycosylase domain by a long spacer. In contrast to UNG, SMUG1 and TDG, which all share the same structural fold, the C-terminal glycosylase domain of MBD4 belongs to the helix-hairpin-helix DNA glycosylase family[129] that includes AlkA, MutY, OGG1 as well as Family 6 UDGs. Nevertheless, MBD4 has overlapping substrate specificity with TDG by removing U, T, 5-hmU and thymine glycol mispaired with G, halogenated pyrimidines and εC (Table 4.1). Mice deficient in MBD4 have increased C to T transition mutations at CpG sites and accelerated intestinal tumour formation.[130,131] Moreover, MBD4 is mutated or inactivated in many human cancers that exhibit microsatellite instability.[132] In contrast to TDG, MBD4 is expressed in all phases of the cell cycle.[121] In addition to its roles in genomic stability, MBD4 has been linked to apoptosis, transcriptional regulation and active demethylation.[116] The possible involvement of MBD4 in cancer development is more extensively discussed in Chapter 7, Sections 7.1 and 7.5.

4.8. UDGs in Hyperthermophilic Organisms and Some Human Pathogens

The rate of cytosine deamination is a function of temperature. Consequently, one would expect to find UDG activity in hyperthermophilic organisms. This was indeed the case in studies on *Bacillus stearothermophilus*[133] and *Thermothrix thiopara*.[134] Except for the thermo-resistance, the purified UDGs from these bacteria had properties similar to those of other known UDGs and the early work has apparently not been followed up. However, DNA sequence analysis of UDG in *B. stearothermophilus* has shown that it encodes an Ung-type enzyme (professor Finn Drabløs, unpublished data). Following the initial papers, UDG activity was found in extracts from seven out of seven hyperthermophilic organisms, among which one bacterium and six archae.[135] To our knowledge, this was the first demonstration of UDG activity in archae. Although the UDGs were not assigned to a family, these enzymes were only partially inhibited (22–87%) by Ugi under condition where *E. coli* Ung was inhibited 100%. Inhibition remained incomplete for the hyperthermophilic UDGs even in the presence of very high concentrations of Ugi, indicating that the UDG enzymes were related to, but different from the Ung protein, or that the extract contained more than one UDG type.

4.8.1. *Family 4 UDGs in Hyperthermophiles and Some Human Pathogens*

The first definite report on a Family 4 member in the UDG superfamily was the report on a thermostable UDG in the anaerobic marine bacterium *Thermotoga maritima*. A gene with very weak homology to *E. coli* T/U-mismatch DNA glycosylase (Mug) was identified by a BLAST search, cloned, expressed and purified. This heat stable UDG removed uracil from DNA both from U:G and U:A contexts, as well as from single stranded DNA. Interestingly, analogous genes were identified in several other prokaryotes, including thermophilic and mesophilic eubacteria and archae, as well as the human pathogens *Treponema pallidum* causing syphilis and *Rickettsia prowazekii* causing epidemic typhus, and the radioresistant organism *Deioncoccus radiodurans*.[15] Although the *T. maritima* Family 4 UDG sequence is not related to *E.coli* Ung, the substrate preference is similar and it can partially substitute for Ung in *E. coli* in BER in cell extracts.[136] The Family 4 UDG from the hyperthermophilic *Archaeglobus fulgidus*, the first one to be isolated from archae, is highly homologous to the enzyme from *T. maritima* and has similar substrate properties. It is even more resistant to high temperature and remains fully active after 1.5 h at 95°C.[137] Interestingly, unlike other UDGs, Family 4 UDG in *Pyrobaculum aerophilum* was found to contain an iron-sulfur cluster [4Fe-4S], and conservation of the four Cys-residues in other thermophilic members of Family 4 UDGs indicates that this is a common feature.[138] Similar to the iron-sulfur centers in Nth/MutY family DNA glycosylases,[139] it is distant from the active site and is thought to have a structural rather than a catalytic role.

4.8.2. *Family 5 UDGs have Surprisingly Broad Substrate Specificity*

The hyperthermophile *P. aerophilum* expresses three different UDG activities, a mismatch glycosylase activity (Pa-MIG) that can remove thymine or uracil from T:G or U:G mismatches, one Family 4 type UDG, as described above, and a Family 5 UDG.[16,138,140] Importantly, *P. aerophilum* also contains an AP-endonuclease, a DNA polymerase and DNA ligase activities that can support complete BER of deaminated cytosines and 5-methylcytosines (5-mC) in DNA.[141] Family 5 UDGs (based on data from *P. aerophilum* and *T. thermophiles*) have much broader substrate specificity than other UDGs, including both damaged or aberrant pyrimidines and purines.[16] Pyrimidine substrates include U:G, U:A, hmU:G, hmU:A, U:T, U:C, εC:G and 5FU:G, but apparently not T:G. The purine substrates include Hx:T, Hx:C, Hx:G, Hx:A (Hx is hypoxanthine), as well as xanthine.[16,142] *Mycobacterium smegmatis* contains both a Family 1 (UdgB) and a Family 5 UDG. While knockout of UdgB resulted in a modest two-fold increased mutation rate, compared with ~8.4-fold increase for the Ung-deficient cell, the Ung/UdgB double knockout displayed ~19.6-fold increased mutation rate, demonstrating that Ung and UdgB complement each other partially.[143]

These findings demonstrate that Family 5 UDGs may initiate repair of uracil from deaminated cytosine, as well as deaminated adenines and guanines. In addition members of this UDG family remove misincorporated bases, although perhaps less efficiently.

4.8.3. *Family 6 UDGs Contain Hairpin-hairpin-helix and a [4Fe-4S] Motif*

The hyperthermophile *Methanococcus jannaschii* has UDG with a helix-hairpin-helix and a [4Fe-4S] motif that, unlike Family 4 UDGs, appear to have a catalytic role. It is a representative of Family 6 UDGs. This enzyme (*Mj*UDG) removes uracil from double-stranded DNA (in pairs or mismatches) and single-stranded DNA. Interestingly, it also removes 8-oxoG (in 8-oxoG:C) from DNA, although less efficiently.

Homologs of this protein were identified in other organisms.[144] Furthermore, the previously identified Pa-MIG in *P. aerophilum* is also related to *Mj*UDG.[145]

4.9. Conclusions

DNA glycosylases that can remove uracil from DNA comprise at least six fairly distinct families that may have a common ancient ancestor. The only possible exception is MBD4 that cannot clearly be assigned to any of these families. While some UDGs, such as Family 1 UDGs, have a very strong specificity for uracil, Family 2 enzymes TDG/Mug and Family 3 enzyme SMUG1 also remove thymine and 5-hmU, respectively. Family 5 enzymes can remove a wide range of modified pyrimidines and purines. Furthermore, the family picture may still not be complete. Thus, the soil bacterium *Bradyrhizobium diazoefficiens* encodes a UDG (*Bdi*UDG) that shows similarity to Family 4 UDGs in its overall fold, but has the Family 1 UDG active site residues, although GQD**PA** (in *Bdi*UDG) compared with GQD**PY** in the same position in classical UNG proteins. This difference may contribute to the broad specificity of *Bdi*UDG.[146] The ubiquitous presence of systems for removal of genomic uracil, including a wide spectrum of DNA glycosylases, as well as mismatch repair, suggest that removal of genomic uracil is important for genomic stability. Furthermore, nature has taken advantage of the conversion of DNA-cytosine to uracil in innate and adaptive immunity, and apparently in epigenetics.

References

1. Lindahl, T., Nyberg, B. Heat-induced deamination of cytosine residues in deoxyribonucleic acid. *Biochemistry* **13**, 3405–3410 (1974).
2. Lindahl, T. An N-glycosidase from Escherichia coli that releases free uracil from DNA containing deaminated cytosine residues. *Proc Natl Acad Sci U S A* **71**, 3649–3653 (1974).
3. Lindahl, T. Instability and decay of the primary structure of DNA. *Nature* **362**, 709–715 (1993).

4. Lindahl, T., Andersson, A. Rate of chain breakage at apurinic sites in double-stranded deoxyribonucleic acid. *Biochemistry* **11**, 3618–3623 (1972).

5. Lindahl, T., Karlstrom, O. Heat-induced depyrimidination of deoxyribonucleic acid in neutral solution. *Biochemistry* **12**, 5151–5154 (1973).

6. Lindahl, T., Nyberg, B. Rate of depurination of native deoxyribonucleic acid. *Biochemistry* **11**, 3610–3618 (1972).

7. Hurst, R. O., Kuksis, A. Degradation of deoxyribonucleic acid by hot alkali. *Can J Biochem Physiol* **36**, 919–929 (1958).

8. Shapiro, R., Klein, R. S. The deamination of cytidine and cytosine by acidic buffer solutions. Mutagenic implications. *Biochemistry* **5**, 2358–2362 (1966).

9. Worrad, D. M., Caradonna, S. Identification of the coding sequence for herpes simplex virus uracil-DNA glycosylase. *J Virol* **62**, 4774–4777 (1988).

10. Varshney, U., Hutcheon, T., van de Sande, J. H. Sequence analysis, expression, and conservation of Escherichia coli uracil DNA glycosylase and its gene (ung). *J Biol Chem* **263**, 7776–7784 (1988).

11. Olsen, L. C., Aasland, R., Wittwer, C. U., *et al.* Molecular cloning of human uracil-DNA glycosylase, a highly conserved DNA repair enzyme. *EMBO J* **8**, 3121–3125 (1989).

12. Brown, T. C., Jiricny, J. A specific mismatch repair event protects mammalian cells from loss of 5-methylcytosine. *Cell* **50**, 945–950 (1987).

13. Neddermann, P., Gallinari, P., Lettieri, T., *et al.* Cloning and expression of human G/T mismatch-specific thymine-DNA glycosylase. *J Biol Chem* **271**, 12767–12774 (1996).

14. Neddermann, P., Jiricny, J. The purification of a mismatch-specific thymine-DNA glycosylase from HeLa cells. *J Biol Chem* **268**, 21218–21224 (1993).

15. Sandigursky, M., Franklin, W. A. Thermostable uracil-DNA glycosylase from Thermotoga maritima a member of a novel class of DNA repair enzymes. *Curr Biol* **9**, 531–534 (1999).

16. Sartori, A. A., Fitz-Gibbon, S., Yang, H., *et al.* A novel uracil-DNA glycosylase with broad substrate specificity and an unusual active site. *EMBO J* **21**, 3182–3191 (2002).

17. Friedberg, E. C., Goldthwait, D. A. Endonuclease II of E. coli. I. Isolation and purification. *Proc Natl Acad Sci U S A* **62**, 934–940 (1969).

18. Hadi, S. M., Goldthwait, D. A. Endonuclease II of Escherichia coli. Degradation of partially depurinated deoxyribonucleic acid. *Biochemistry* **10**, 4986–4993 (1971).

19. Lindahl, T. My journey to DNA repair. *Genomics Proteomics Bioinformatics* **11**, 2–7 (2013).

20. Lindahl, T. New class of enzymes acting on damaged DNA. *Nature* **259**, 64–66 (1976).

21. Karran, P., Lindahl, T. Enzymatic excision of free hypoxanthine from poly-deoxynucleotides and DNA containing deoxyinosine monophosphate residues. *J Biol Chem* **253**, 5877–5879 (1978).
22. Chetsanga, C. J., Lindahl, T. Release of 7-methylguanine residues whose imidazole rings have been opened from damaged DNA by a DNA glycosylase from Escherichia coli. *Nucleic Acids Res* **6**, 3673–3684 (1979).
23. Krokan, H. E., Bjoras, M. Base excision repair. *Cold Spring Harb Perspect Biol* **5**, a012583 (2013).
24. Lindahl, T., Ljungquist, S., Siegert, W., *et al.* DNA N-glycosidases: properties of uracil-DNA glycosidase from Escherichia coli. *J Biol Chem* **252**, 3286–3294 (1977).
25. Tye, B. K., Lehman, I. R. Excision repair of uracil incorporated in DNA as a result of a defect in dUTPase. *J Mol Biol* **117**, 293–306 (1977).
26. Tye, B. K., Nyman, P. O., Lehman, I. R., *et al.* Transient accumulation of Okazaki fragments as a result of uracil incorporation into nascent DNA. *Proc Natl Acad Sci U S A* **74**, 154–157 (1977).
27. Grafstrom, R. H., Tseng, B. Y., Goulian, M. The incorporation of uracil into animal cell DNA in vitro. *Cell* **15**, 131–140 (1978).
28. Wist, E., Unhjem, O., Krokan, H. Accumulation of small fragments of DNA in isolated HeLa cell nuclei due to transient incorporation of dUMP. *Biochim Biophys Acta* **520**, 253–270 (1978).
29. Krokan, H. E., Drablos, F., Slupphaug, G. Uracil in DNA--occurrence, consequences and repair. *Oncogene* **21**, 8935–8948 (2002).
30. Cone, R., Duncan, J., Hamilton, L., Friedberg, E. C. Partial purification and characterization of a uracil DNA N-glycosidase from Bacillus subtilis. *Biochemistry* **16**, 3194–3201 (1977).
31. Tomlin, N. V., Aprelikova, O. N., Barenfeld, L. S. Enzymes from Micrococcus luteus involved in the initial steps of excision repair of spontaneous DNA lesions: uracil-DNA-glycosidase and apurinic-endonucleases. *Nucleic Acids Res* **5**, 1413–1428 (1978).
32. Caradonna, S. J., Cheng, Y. C. Induction of uracil-DNA glycosylase and dUTP nucleotidohydrolase activity in herpes simplex virus-infected human cells. *J Biol Chem* **256**, 9834–9837 (1981).
33. Stuart, D. T., Upton, C., Higman, M. A., *et al.* A poxvirus-encoded uracil DNA glycosylase is essential for virus viability. *J Virol* **67**, 2503–2512 (1993).
34. Crosby, B., Prakash, L., Davis, H., Hinkle, D. C. Purification and characterization of a uracil-DNA glycosylase from the yeast. Saccharomyces cerevisiae. *Nucleic Acids Res* **9**, 5797–5809 (1981).
35. Birch, D. J., McLennan, A. G. Uracil-DNA glycosylase in developing embryos of the brine shrimp (Artemia salina). *Biochem Soc Trans* **8**, 730–731 (1980).

36. Lanes, O., Guddal, P. H., Gjellesvik, D. R., Willassen, N. P. Purification and characterization of a cold-adapted uracil-DNA glycosylase from Atlantic cod (Gadus morhua). *Comp Biochem Physiol B Biochem Mol Biol* **127**, 399–410 (2000).

37. Blaisdell, P., Warner, H. Partial purification and characterization of a uracil-DNA glycosylase from wheat germ. *J Biol Chem* **258**, 1603–1609 (1983).

38. Talpaert-Borle, M., Liuzzi, M. Base-excision repair in carrot cells. Partial purification and characterization of uracil-DNA glycosylase and apurinic/apyrimidinic endodeoxyribonuclease. *Eur J Biochem* **124**, 435–440 (1982).

39. Makino, F., Munakata, N. Excision of uracil from bromodeoxyuridine-substituted and U.V.-irradiated DNA in cultured mouse lymphoma cells. *Int J Radiat Biol Relat Stud Phys Chem Med* **36**, 349–357 (1979).

40. Sekiguchi, M., Hayakawa, H., Makino, F., *et al.* A human enzyme that liberates uracil from DNA. *Biochem Biophys Res Commun* **73**, 293–299 (1976).

41. Kuhnlein, U., Lee, B., Linn, S. Human uracil DNA N-glycosidase: studies in normal and repair defective cultured fibroblasts. *Nucleic Acids Res* **5**, 117–125 (1978).

42. Lucas-Lledo, J. I., Maddamsetti, R., Lynch, M. Phylogenomic analysis of the uracil-DNA glycosylase superfamily. *Mol Biol Evol* **28**, 1307–1317 (2011).

43. Schormann, N., Ricciardi, R., Chattopadhyay, D. Uracil-DNA glycosylases-structural and functional perspectives on an essential family of DNA repair enzymes. *Protein Sci* **23**, 1667–1685 (2014).

44. Duncan, B. K., Rockstroh, P. A., Warner, H. R. Escherichia coli K-12 mutants deficient in uracil-DNA glycosylase. *J Bacteriol* **134**, 1039–1045 (1978).

45. Duncan, B. K., Weiss, B. Specific mutator effects of ung (uracil-DNA glycosylase) mutations in Escherichia coli. *J Bacteriol* **151**, 750–755 (1982).

46. Lutsenko, E., Bhagwat, A. S. The role of the Escherichia coli mug protein in the removal of uracil and 3,N(4)-ethenocytosine from DNA. *J Biol Chem* **274**, 31034–31038 (1999).

47. Gallinari, P., Jiricny, J. A new class of uracil-DNA glycosylases related to human thymine-DNA glycosylase. *Nature* **383**, 735–738 (1996).

48. Bolt, H. M. Vinyl chloride-a classical industrial toxicant of new interest. *Crit Rev Toxicol* **35**, 307–323 (2005).

49. Mokkapati, S. K., Fernandez de Henestrosa, A. R., Bhagwat, A. S. Escherichia coli DNA glycosylase Mug: a growth-regulated enzyme required for mutation avoidance in stationary-phase cells. *Mol Microbiol* **41**, 1101–1111 (2001).

50. Krokan, H., Wittwer, C. U. Uracil DNa-glycosylase from HeLa cells: general properties, substrate specificity and effect of uracil analogs. *Nucleic Acids Res* **9**, 2599–2613 (1981).

51. Wittwer, C. U., Krokan, H. Uracil-DNA glycosylase in HeLa S3 cells: interconvertibility of 50 and 20 kDa forms and similarity of the nuclear and mitochondrial form of the enzyme. *Biochim Biophys Acta* **832**, 308–318 (1985).

52. Anderson, C. T., Friedberg, E. C. The presence of nuclear and mitochondrial uracil-DNA glycosylase in extracts of human KB cells. *Nucleic Acids Res* **8**, 875–888 (1980).

53. Gupta, P. K., Sirover, M. A. Stimulation of the nuclear uracil DNA glycosylase in proliferating human fibroblasts. *Cancer Res* **41**, 3133–3136 (1981).

54. Caradonna, S. J., Cheng, Y. C. Uracil DNA-glycosylase. Purification and properties of this enzyme isolated from blast cells of acute myelocytic leukemia patients. *J Biol Chem* **255**, 2293–2300 (1980).

55. Myrnes, B., Wittwer, C. U. Purification of the human O6-methylguanine-DNA methyltransferase and uracil-DNA glycosylase, the latter to apparent homogeneity. *Eur J Biochem* **173**, 383–387 (1988).

56. Meyer-Siegler, K., Mauro, D. J., Seal, G., *et al.* A human nuclear uracil DNA glycosylase is the 37-kDa subunit of glyceraldehyde-3-phosphate dehydrogenase. *Proc Natl Acad Sci U S A* **88**, 8460–8464 (1991).

57. Arenaz, P., Sirover, M. A. Isolation and characterization of monoclonal antibodies directed against the DNA repair enzyme uracil DNA glycosylase from human placenta. *Proc Natl Acad Sci U S A* **80**, 5822–5826 (1983).

58. Muller, S. J., Caradonna, S. Isolation and characterization of a human cDNA encoding uracil-DNA glycosylase. *Biochim Biophys Acta* **1088**, 197–207 (1991).

59. Muller, S. J., Caradonna, S. Cell cycle regulation of a human cyclin-like gene encoding uracil-DNA glycosylase. *J Biol Chem* **268**, 1310–1319 (1993).

60. Wittwer, C. U., Bauw, G., Krokan, H. E. Purification and determination of the NH2-terminal amino acid sequence of uracil-DNA glycosylase from human placenta. *Biochemistry* **28**, 780–784 (1989).

61. Olsen, L. C., Aasland, R., Krokan, H. E., Helland, D. E. Human uracil-DNA glycosylase complements E. coli ung mutants. *Nucleic Acids Res* **19**, 4473–4478 (1991).

62. Davison, A. J., Scott, J. E. The complete DNA sequence of varicella-zoster virus. *J Gen Virol* **67** (**Pt 9**), 1759–1816 (1986).

63. Upton, C., Stuart, D. T., McFadden, G. Identification of a poxvirus gene encoding a uracil DNA glycosylase. *Proc Natl Acad Sci U S A* **90**, 4518–4522 (1993).

64. Percival, K. J., Klein, M. B., Burgers, P. M. Molecular cloning and primary structure of the uracil-DNA-glycosylase gene from Saccharomyces cerevisiae. *J Biol Chem* **264**, 2593–2598 (1989).

65. Karran, P., Cone, R., Friedberg, E. C. Specificity of the bacteriophage PBS2 induced inhibitor of uracil-DNA glycosylase. *Biochemistry* **20**, 6092–6096 (1981).

66. Haug, T., Skorpen, F., Kvaloy, K., *et al.* Human uracil-DNA glycosylase gene: sequence organization, methylation pattern, and mapping to chromosome 12q23-q24.1. *Genomics* **36**, 408–416 (1996).

67. Nilsen, H., Otterlei, M., Haug, T., *et al.* Nuclear and mitochondrial uracil-DNA glycosylases are generated by alternative splicing and transcription from different positions in the UNG gene. *Nucleic Acids Res* **25**, 750–755 (1997).

68. Kavli, B., Sundheim, O., Akbari, M., *et al.* hUNG2 is the major repair enzyme for removal of uracil from U:A matches, U:G mismatches, and U in single-stranded DNA, with hSMUG1 as a broad specificity backup. *J Biol Chem* **277**, 39926–39936 (2002).

69. Doseth, B., Visnes, T., Wallenius, A., *et al.* Uracil-DNA glycosylase in base excision repair and adaptive immunity: species differences between man and mouse. *J Biol Chem* **286**, 16669–16680 (2011).

70. Bharati, S., Krokan, H. E., Kristiansen, L., *et al.* Human mitochondrial uracil-DNA glycosylase preform (UNG1) is processed to two forms one of which is resistant to inhibition by AP sites. *Nucleic Acids Res* **26**, 4953–4959 (1998).

71. Nilsen, H., Rosewell, I., Robins, P., *et al.* Uracil-DNA glycosylase (UNG)-deficient mice reveal a primary role of the enzyme during DNA replication. *Mol Cell* **5**, 1059–1065 (2000).

72. Otterlei, M., Warbrick, E., Nagelhus, T. A., *et al.* Post-replicative base excision repair in replication foci. *EMBO J* **18**, 3834–3844 (1999).

73. Hagen, L., Kavli, B., Sousa, M. M., *et al.* Cell cycle-specific UNG2 phosphorylations regulate protein turnover, activity and association with RPA. *EMBO J* **27**, 51–61 (2008).

74. Muramatsu, M., Kinoshita, K., Fagarasan, S., *et al.* Class switch recombination and hypermutation require activation-induced cytidine deaminase (AID), a potential RNA editing enzyme. *Cell* **102**, 553–563 (2000).

75. Revy, P., Muto, T., Levy, Y., *et al.* Activation-induced cytidine deaminase (AID) deficiency causes the autosomal recessive form of the Hyper-IgM syndrome (HIGM2). *Cell* **102**, 565–575 (2000).

76. Maul, R. W., Gearhart, P. J. Refining the Neuberger model: Uracil processing by activated B cells. *Eur J Immunol* **44**, 1913–1916 (2014).

77. Dizdaroglu, M., Karakaya, A., Jaruga, P., *et al.* Novel activities of human uracil DNA N-glycosylase for cytosine-derived products of oxidative DNA damage. *Nucleic Acids Res* **24**, 418–422 (1996).

78. Xue, J. H., Xu, G. F., Gu, T. P., *et al.* Uracil-DNA glycosylase UNG promotes Tet-mediated DNA demethylation. *J Biol Chem* (2015).

79. Akbari, M., Otterlei, M., Pena-Diaz, J., Krokan, H. E. Different organization of base excision repair of uracil in DNA in nuclei and mitochondria and selective upregulation of mitochondrial uracil-DNA glycosylase after oxidative stress. *Neuroscience* **145**, 1201–1212 (2007).
80. Otterlei, M., Haug, T., Nagelhus, T. A., *et al.* Nuclear and mitochondrial splice forms of human uracil-DNA glycosylase contain a complex nuclear localisation signal and a strong classical mitochondrial localisation signal, respectively. *Nucleic Acids Res* **26**, 4611–4617 (1998).
81. Akbari, M., Otterlei, M., Pena-Diaz, J., *et al.* Repair of U/G and U/A in DNA by UNG2-associated repair complexes takes place predominantly by short-patch repair both in proliferating and growth-arrested cells. *Nucleic Acids Res* **32**, 5486–5498 (2004).
82. Akbari, M., Solvang-Garten, K., Hanssen-Bauer, A., *et al.* Direct interaction between XRCC1 and UNG2 facilitates rapid repair of uracil in DNA by XRCC1 complexes. *DNA Repair (Amst)* **9**, 785–795 (2010).
83. Hanssen-Bauer, A., Solvang-Garten, K., Sundheim, O., Pena-Diaz, J., Andersen, S., Slupphaug, G., Krokan, H. E., Wilson, D. M., 3rd, Akbari, M., Otterlei, M. XRCC1 coordinates disparate responses and multiprotein repair complexes depending on the nature and context of the DNA damage. *Environ Mol Mutagen* **52**, 623–635 (2011).
84. Nagelhus, T. A., Haug, T., Singh, K. K., *et al.* A sequence in the N-terminal region of human uracil-DNA glycosylase with homology to XPA interacts with the C-terminal part of the 34-kDa subunit of replication protein A. *J Biol Chem* **272**, 6561–6566 (1997).
85. Mol, C. D., Arvai, A. S., Slupphaug, G., *et al.* Crystal structure and mutational analysis of human uracil-DNA glycosylase: structural basis for specificity and catalysis. *Cell* **80**, 869–878 (1995).
86. Slupphaug, G., Mol, C. D., Kavli, B., *et al.* A nucleotide-flipping mechanism from the structure of human uracil-DNA glycosylase bound to DNA. *Nature* **384**, 87–92 (1996).
87. Savva, R., McAuley-Hecht, K., Brown, T., Pearl, L. The structural basis of specific base-excision repair by uracil-DNA glycosylase. *Nature* **373**, 487–493 (1995).
88. Kavli, B., Slupphaug, G., Mol, C. D., *et al.* Excision of cytosine and thymine from DNA by mutants of human uracil-DNA glycosylase. *EMBO J* **15**, 3442–3447 (1996).
89. Otterlei, M., Kavli, B., Standal, R., *et al.* Repair of chromosomal abasic sites in vivo involves at least three different repair pathways. *EMBO J* **19**, 5542–5551 (2000).

90. Auerbach, P., Bennett, R. A., Bailey, E. A., *et al.* Mutagenic specificity of endogenously generated abasic sites in Saccharomyces cerevisiae chromosomal DNA. *Proc Natl Acad Sci U S A* **102**, 17711–17716 (2005).

91. Lauritzen, K. H., Cheng, C., Wiksen, H., *et al.* Mitochondrial DNA toxicity compromises mitochondrial dynamics and induces hippocampal antioxidant defenses. *DNA Repair (Amst)* **10**, 639–653 (2011).

92. Lauritzen, K. H., Moldestad, O., Eide, L., *et al.* Mitochondrial DNA toxicity in forebrain neurons causes apoptosis, neurodegeneration, and impaired behavior. *Mol Cell Biol* **30**, 1357–1367 (2010).

93. Lauritzen, K. H., Kleppa, L., Aronsen, J. M., *et al.* Impaired dynamics and function of mitochondria caused by mtDNA toxicity leads to heart failure. *Am J Physiol Heart Circ Physiol* **309**, H434–449 (2015).

94. Parikh, S. S., Mol, C. D., Slupphaug, G., *et al.* Base excision repair initiation revealed by crystal structures and binding kinetics of human uracil-DNA glycosylase with DNA. *EMBO J* **17**, 5214–5226 (1998).

95. Parikh, S. S., Walcher, G., Jones, G. D., *et al.* Uracil-DNA glycosylase-DNA substrate and product structures: conformational strain promotes catalytic efficiency by coupled stereoelectronic effects. *Proc Natl Acad Sci USA* **97**, 5083–5088 (2000).

96. Cao, C., Jiang, Y. L., Krosky, D. J., Stivers, J. T. The catalytic power of uracil DNA glycosylase in the opening of thymine base pairs. *J Am Chem Soc* **128**, 13034–13035 (2006).

97. Friedman, J. I., Stivers, J. T. Detection of damaged DNA bases by DNA glycosylase enzymes. *Biochemistry* **49**, 4957–4967 (2010).

98. Huffman, J. L., Sundheim, O., Tainer, J. A. DNA base damage recognition and removal: new twists and grooves. *Mutat Res* **577**, 55–76 (2005).

99. Krokan, H. E., Standal, R., Slupphaug, G. DNA glycosylases in the base excision repair of DNA. *Biochem J* **325** (Pt 1), 1–16 (1997).

100. Klimasauskas, S., Kumar, S., Roberts, R. J., Cheng, X. HhaI methyltransferase flips its target base out of the DNA helix. *Cell* **76**, 357–369 (1994).

101. Endres, M., Biniszkiewicz, D., Sobol, R. W., *et al.* Increased postischemic brain injury in mice deficient in uracil-DNA glycosylase. *J Clin Invest* **113**, 1711–1721 (2004).

102. Rada, C., Williams, G. T., Nilsen, H., *et al.* Immunoglobulin isotype switching is inhibited and somatic hypermutation perturbed in UNG-deficient mice. *Curr Biol* **12**, 1748–1755 (2002).

103. Andersen, S., Ericsson, M., Dai, H. Y., *et al.* Monoclonal B-cell hyperplasia and leukocyte imbalance precede development of B-cell malignancies in uracil-DNA glycosylase deficient mice. *DNA Repair (Amst)* **4**, 1432–1441 (2005).

104. Nilsen, H., Stamp, G., Andersen, S., *et al.* Gene-targeted mice lacking the Ung uracil-DNA glycosylase develop B-cell lymphomas. *Oncogene* **22**, 5381–5386 (2003).

105. Andersen, S., Heine, T., Sneve, R., *et al.* Incorporation of dUMP into DNA is a major source of spontaneous DNA damage, while excision of uracil is not required for cytotoxicity of fluoropyrimidines in mouse embryonic fibroblasts. *Carcinogenesis* **26**, 547–555 (2005).

106. Galashevskaya, A., Sarno, A., Vagbo, C. B., *et al.* A robust, sensitive assay for genomic uracil determination by LC/MS/MS reveals lower levels than previously reported. *DNA Repair (Amst)* **12**, 699–706 (2013).

107. Imai, K., Slupphaug, G., Lee, W. I., *et al.* Human uracil-DNA glycosylase deficiency associated with profoundly impaired immunoglobulin class-switch recombination. *Nat Immunol* **4**, 1023–1028 (2003).

108. Kronenberg, G., Harms, C., Sobol, R. W., *et al.* Folate deficiency induces neurodegeneration and brain dysfunction in mice lacking uracil DNA glycosylase. *J Neurosci* **28**, 7219–7230 (2008).

109. Kemmerich, K., Dingler, F. A., Rada, C., Neuberger, M. S. Germline ablation of SMUG1 DNA glycosylase causes loss of 5-hydroxymethyluracil- and UNG-backup uracil-excision activities and increases cancer predisposition of Ung-/-Msh2-/- mice. *Nucleic Acids Res* **40**, 6016–6025 (2012).

110. Zeitlin, S. G., Chapados, B. R., Baker, N. M., *et al.* Uracil DNA N-glycosylase promotes assembly of human centromere protein A. *PLoS One* **6**, e17151 (2011).

111. Zeitlin, S. G., Patel, S., Kavli, B., Slupphaug, G. Xenopus CENP-A assembly into chromatin requires base excision repair proteins. *DNA Repair (Amst)* **4**, 760–772 (2005).

112. Baba, D., Maita, N., Jee, J. G., *et al.* Crystal structure of thymine DNA glycosylase conjugated to SUMO-1. *Nature* **435**, 979–982 (2005).

113. Hardeland, U., Bentele, M., Lettieri, T., *et al.* Thymine DNA glycosylase. *Prog Nucleic Acid Res Mol Biol* **68**, 235–253 (2001).

114. Hardeland, U., Kunz, C., Focke, F., *et al.* Cell cycle regulation as a mechanism for functional separation of the apparently redundant uracil DNA glycosylases TDG and UNG2. *Nucleic Acids Res* **35**, 3859–3867 (2007).

115. Cortazar, D., Kunz, C., Selfridge, J., *et al.* Embryonic lethal phenotype reveals a function of TDG in maintaining epigenetic stability. *Nature* **470**, 419–423 (2011).

116. Sjolund, A. B., Senejani, A. G., Sweasy, J. B. MBD4 and TDG: multifaceted DNA glycosylases with ever expanding biological roles. *Mutat Res* **743–744**, 12–25 (2013).

117. He, Y. F., Li, B. Z., Li, Z., *et al.* Tet-mediated formation of 5-carboxylcytosine and its excision by TDG in mammalian DNA. *Science* **333**, 1303–1307 (2011).

118. Maiti, A., Drohat, A. C. Thymine DNA glycosylase can rapidly excise 5-formyl-cytosine and 5-carboxylcytosine: potential implications for active demethylation of CpG sites. *J Biol Chem* **286**, 35334–35338 (2011).

119. Haushalter, K. A., Todd Stukenberg, M. W., Kirschner, M. W., Verdine, G. L. Identification of a new uracil-DNA glycosylase family by expression cloning using synthetic inhibitors. *Curr Biol* **9**, 174–185 (1999).

120. Pearl, L. H. Structure and function in the uracil-DNA glycosylase superfamily. *Mutat Res* **460**, 165–181 (2000).

121. Mjelle, R., Hegre, S. A., Aas, P. A., *et al.* Cell cycle regulation of human DNA repair and chromatin remodeling genes. *DNA Repair (Amst)* **30**, 53–67 (2015).

122. Pettersen, H. S., Sundheim, O., Gilljam, K. M., *et al.* Uracil-DNA glycosylases SMUG1 and UNG2 coordinate the initial steps of base excision repair by distinct mechanisms. *Nucleic Acids Res* **35**, 3879–3892 (2007).

123. Doseth, B., Ekre, C., Slupphaug, G., *et al.* Strikingly different properties of uracil-DNA glycosylases UNG2 and SMUG1 may explain divergent roles in processing of genomic uracil. *DNA Repair (Amst)* **11**, 587–593 (2012).

124. Nilsen, H., Haushalter, K. A., Robins, P., *et al.* Excision of deaminated cytosine from the vertebrate genome: role of the SMUG1 uracil-DNA glycosylase. *EMBO J* **20**, 4278–4286 (2001).

125. Jobert, L., Skjeldam, H. K., Dalhus, B., *et al.* The human base excision repair enzyme SMUG1 directly interacts with DKC1 and contributes to RNA quality control. *Mol Cell* **49**, 339–345 (2013).

126. Alsoe, L., Sarno, A., Carracedo, S., *et al.* Uracil Accumulation and Mutagenesis Dominated by Cytosine Deamination in CpG Dinucleotides in Mice Lacking UNG and SMUG1. *Sci Rep* **7**, 7199 (2017).

127. Boorstein, R. J., Cummings, A., Jr., Marenstein, D. R., *et al.* Definitive identification of mammalian 5-hydroxymethyluracil DNA N-glycosylase activity as SMUG1. *J Biol Chem* **276**, 41991–41997 (2001).

128. Pettersen, H. S., Visnes, T., Vagbo, C. B., *et al.* UNG-initiated base excision repair is the major repair route for 5-fluorouracil in DNA, but 5-fluorouracil cytotoxicity depends mainly on RNA incorporation. *Nucleic Acids Res* **39**, 8430–8444 (2011).

129. Wu, P., Qiu, C., Sohail, A., *et al.* Mismatch repair in methylated DNA. Structure and activity of the mismatch-specific thymine glycosylase domain of methyl-CpG-binding protein MBD4. *J Biol Chem* **278**, 5285–5291 (2003).

130. Millar, C. B., Guy, J., Sansom, O. J., *et al.* Enhanced CpG mutability and tumorigenesis in MBD4-deficient mice. *Science* **297**, 403–405 (2002).

131. Wong, E., Yang, K., Kuraguchi, M., *et al.* Mbd4 inactivation increases Cright-arrowT transition mutations and promotes gastrointestinal tumor formation. *Proc Natl Acad Sci U S A* **99**, 14937–14942 (2002).

132. Tricarico, R., Cortellino, S., Riccio, A., *et al.* Involvement of MBD4 inactivation in mismatch repair-deficient tumorigenesis. *Oncotarget* (2015).

133. Kaboev, O. K., Luchkina, L. A., Akhmedov, A. T., Bekker, M. L. Uracil-DNA glycosylase from Bacillus stearothermophilus. *FEBS Lett* **132**, 337–340 (1981).

134. Kaboev, O. K., Luchkina, L. A., Kuziakina, T. I. Uracil-DNA glycosylase of thermophilic Thermothrix thiopara. *J Bacteriol* **164**, 421–424 (1985).

135. Koulis, A., Cowan, D. A., Pearl, L. H., Savva, R. Uracil-DNA glycosylase activities in hyperthermophilic micro-organisms. *FEMS Microbiol Lett* **143**, 267–271 (1996).

136. Sandigursky, M., Faje, A., Franklin, W. A. Characterization of the full length uracil-DNA glycosylase in the extreme thermophile Thermotoga maritima. *Mutat Res* **485**, 187–195 (2001).

137. Sandigursky, M., Franklin, W. A. Uracil-DNA glycosylase in the extreme thermophile Archaeoglobus fulgidus. *J Biol Chem* **275**, 19146–19149 (2000).

138. Hinks, J. A., Evans, M. C., De Miguel, Y., *et al.* An iron-sulfur cluster in the family 4 uracil-DNA glycosylases. *J Biol Chem* **277**, 16936–16940 (2002).

139. Kuo, C. F., McRee, D. E., Fisher, C. L., *et al.* Atomic structure of the DNA repair [4Fe-4S] enzyme endonuclease III. *Science* **258**, 434–440 (1992).

140. Sartori, A. A., Schar, P., Fitz-Gibbon, S., *et al.* Biochemical characterization of uracil processing activities in the hyperthermophilic archaeon Pyrobaculum aerophilum. *J Biol Chem* **276**, 29979–29986 (2001).

141. Sartori, A. A., Jiricny, J. Enzymology of base excision repair in the hyperthermophilic archaeon Pyrobaculum aerophilum. *J Biol Chem* **278**, 24563–24576 (2003).

142. Xia, B., Liu, Y., Li, W., *et al.* Specificity and catalytic mechanism in family 5 uracil DNA glycosylase. *J Biol Chem* **289**, 18413–18426 (2014).

143. Malshetty, V. S., Jain, R., Srinath, T., *et al.* Synergistic effects of UdgB and Ung in mutation prevention and protection against commonly encountered DNA damaging agents in Mycobacterium smegmatis. *Microbiology* **156**, 940–949 (2010).

144. Chung, J. H., Im, E. K., Park, H. Y., *et al.* A novel uracil-DNA glycosylase family related to the helix-hairpin-helix DNA glycosylase superfamily. *Nucleic Acids Res* **31**, 2045–2055 (2003).

145. Yang, H., Fitz-Gibbon, S., Marcotte, E. M., *et al.* Characterization of a thermostable DNA glycosylase specific for U/G and T/G mismatches from the hyperthermophilic archaeon Pyrobaculum aerophilum. *J Bacteriol* **182**, 1272–1279 (2000).
146. Chembazhi, U. V., Patil, V. V., Sah, S., *et al.* Uracil DNA glycosylase (UDG) activities in Bradyrhizobium diazoefficiens: characterization of a new class of UDG with broad substrate specificity. *Nucleic Acids Res* **45**, 5863–5876 (2017).

Chapter 5

Viral Uracil — Uracil DNA Glycosylases and dUTPases

Hans E. Krokan*, Bodil Kavli and Geir Slupphaug

DNA viruses and retroviruses use various strategies to avoid and remove genomic uracil. Similar to the case for cells, dUTPase and UNG-type uracil-DNA glycosylase are central in these processes. Large DNA viruses encode viral variants of Family 1 (UNG) enzymes, while smaller viruses depend solely on host cell enzymes. Genes for viral dUTPase and UNG are among those that are expressed early after infection. *Herpesviridae* and *poxviridae* encode their own uracil-DNA glycosylase, dUTPase and DNA polymerase. Midsize DNA viruses in the *adenoviridae* family encode their own DNA polymerase, but not uracil-DNA glycosylase. Fowl adenovirus 9 also encodes a dUTPase,[1] but in human adenoviruses a dUTPase-like gene has diverged to become a transforming factor named E4-ORF1.[2,3] *Retroviridae* do not encode uracil-DNA glycosylase, but some of them capture host uracil-DNA glycosylase, which is encapsulated in the virus particle and released after infection. Furthermore, several major retroviruses encode dUTPase, although this is not the case for primate lentiviruses that instead use captured UNG to minimize genomic uracil in replicative DNA intermediates.[4] Herpesvirus uracil-DNA glycosylase has strong homology to *E. coli*, yeast and mammalian UNG proteins, while Poxvirus uracil-DNA

*hans.krokan@ntnu.no

glycosylase is a more distant member of the same family. Herpesviruses and adenoviruses replicate in the host nucleus, where they in part can rely on host factors. In contrast, poxviruses replicate in replication factories in the cytosol and encode several DNA replication and repair proteins as well as nucleotide metabolizing enzymes. Interestingly, while herpesvirus and poxvirus UNG proteins are genuine uracil-DNA glycosylases, they also have important functions in viral replication. This has been extensively studied for poxvirus UNG, which is essential for virus replication. Poxvirus UNG (D4) forms a heterotrimeric complex with an adapter protein (A20) and the pox DNA polymerase (E9) and acts as a polymerase processivity factor. Herpes UNG is particularly important for virulence in nerve cells, which lack or have very low levels of cellular UNG. The function of host UNG in *retroviridae* is complex, as outlined below. The special functions of viral UNG and dUTPase make these proteins potential drug targets. The giant mimivirus *Acanthamoeba polyphaga* (see Chapter 2) encodes numerous DNA repair proteins, including an UNG-type uracil-DNA glycosylase. Its sequence and crystal structure demonstrate the presence of a catalytic domain containing structural and functional motifs typically present in other UNG proteins and an unstructured N-terminal extension typical of mammalian UNG proteins.[5]

5.1. Uracil-DNA Glycosylases and dUTPases of *Herpesviridae*

Herpesviridae are double stranded DNA viruses with numerous subfamilies causing disease in different species. Nine human herpesviruses (HHVs) are known. These are herpes simplex virus type 1 and 2 (HSV-1 and 2 or HHV-1 and 2), varicella-zoster virus (VZV or HHV-3), Epstein-Barr virus (EBV or HHV-4), human cytomegalovirus (HCMV or HHV-5), HHV-6A and 6B, HHV-7 and Kaposi's sarcoma-associated herpesvirus. Herpesviruses cause a wide variety of human disease, including skin infections, neurological and developmental disease, immunodeficiency, autoimmune disease and cancer.[6-8] Genomes of HHVs have sizes from approximately 124 (VZV) to 230 (HCMV) million base pairs and have 70–200 open reading frames (ORFs) in two unique sequence regions (UL and US) flanked by repeat regions. The number of encoded proteins is considerably larger due to alternative splicing. In addition, herpesvirus genomes encode several non-coding RNA species with various regulatory functions.[8,9] The mechanisms of replication of herpesviruses are largely conserved. It is believed to start at one of three origins of replication. For herpes simplex virus, seven essential virus-encoded replication proteins are known. In addition, six nonessential auxiliary viral factors have been identified, including viral UNG (UL2) and dUTPase (UL50). It also appears that host UNG and dUTPase can be used, in some cases interchangeably with virus proteins, depending on the status of the cell (replicating or quiescent). A rolling circle model for replication has been proposed, based on the observation that DNA concatemers are intermediates in replication, but the mechanism is apparently more complex and involves recombination-dependent replication.[10] After a primary infection, herpesviruses commonly enter a state of latency in host cells, during which the genomes are functional, but few or no viral particles are formed. Different types of stress may trigger new outbursts of the viral disease. HSV typically reside in sensory cranial nerves during latency, VZV in cranial nerves and dorsal root ganglia of spinal nerves, whereas EBV reside in memory B-cells.[8]

5.1.1. *Herpesvirus UNG and dUTPase are Required for Virulence and Reactivation from Latency*

Herpesvirus UNG and dUTPase are important for reactivation of herpesvirus from latency, thus making these proteins potential new drug targets. The first evidence of a herpesvirus encoded dUTPase and uracil-DNA glycosylase was obtained after infection of HeLa cells with HSV-1 and HSV2. Several-fold increases in uracil-DNA glycosylase and dUTPase activities were observed and found to be distinct from the corresponding host cell activities by electrophoretic mobility and/or immunoreactivity.[11] Similar to the host cell dUTPase, HSV-1 dUTPase hydrolyses dUTP to dUMP and pyrophosphate. It requires divalent cations for activity, preferentially Mg^{2+}, and is inhibited by chelating agents like EDTA. However, it can be distinguished from host cell dUTPase by its isoelectric point. Furthermore, the viral dUTPase is a monomeric protein of 35,000 Da, whereas host cell dUTPase was found to be a homodimer with subunits of 22,500 Da.[12] HSV-1 dUTPase is encoded by the ORF UL50 and is not essential for virus growth in cell culture.[13] However, mutants in HSV-1 dUTPase were later found to be strongly attenuated in neuro-virulence, neuro-invasiveness and reactivation from latency in mouse.[14] Similarly, mutants in HSV-1 UNG displayed dramatically decreased virus replication and neuro-invasiveness in the mouse.[15] These observations strongly indicate that HSV-1 dUTPase and UNG are both required for the virulence *in vivo*, at least in the mouse. This may largely be due to low levels of host UNG-activity in mouse nervous tissue, in contrast to proliferating cultured cells. However, even in proliferating cells HSV-1 dUTPase and UNG are anti-mutator enzymes, although this was only observed after successive passages for UNG-mutants.[16]

HSV-2 uracil-DNA glycosylase is encoded by ORF UL2, as verified by *in vitro* transcription and translation and activity measurements. The molecular weight of uracil-DNA glycosylase predicted by the ORF was 37,700 Da, in reasonable agreement with the size of the purified protein.[17] Similarly, the UL2 of HSV1 was also demonstrated to encode uracil-DNA glycosylase activity.[18] HSV-1 and 2 UNGs display strong similarity to *E. coli* Ung.[19] An ORF of 765 bps in HHV-6 also encodes an active UNG-type uracil-DNA glycosylase, but the activity was apparently relatively low. Unlike active site residues GQDPY that are present in human, *E. coli*, yeast and herpesviruses HSV-1, HSV2, VZV, EBV and HCMV, the active site of HHV-6 contains GHDPY. The overall identity of HHV-6 with other herpesviruses at the protein level is also relatively low, ranging from 13% (HSV-1, HSV-2) to 30% (EBV and HCMV).

5.1.2. *Human Cytomegalovirus UNG Interacts with Replication Proteins*

Human cytomegalovirus (HCMV) infection of pregnant women may cause congenital birth defects and serious disease in immunocompromised individuals.[20] HCMV UNG is among the best studied herpesvirus UNG enzymes both functionally and enzymatically. It is encoded by ORF UL114 and is a highly conserved member of the UNG family. HCMV deleted in ORF UL114 retains capacity to proliferate in quiescent primary human fibroblasts, but the replication is 24–48 h delayed compared with the wild-type.[21] HCMV replicates in subnuclear compartments that contain several presumed viral replication factors, including the polymerase processivity factor UL44. Deletion of UL114 causes delayed formation of replication compartments in quiescent fibroblasts.[22] However, in actively growing fibroblasts that express higher levels of host UNG, such delay was not observed, perhaps because the host cell UNG can complement the lack of viral UNG.[23] Interestingly, a direct interaction between HCMV UNG (U114) and the processivity factor UL44 was observed and found to increase the efficiency of both early and late viral DNA synthesis.[24] *In vitro*, the uracil-DNA glycosylase activity of purified HCMV UNG is 350–650-fold lower than that of human UNG. Although the activity was stimulated some 5-fold by equimolar concentrations of UL44, UL114 remains an inefficient DNA glycosylase, at least *in vitro*. Co-precipitation of UL114 and U44 from cell extracts, and the presence of UL114 in replication compartments support the concept of UL114 as a replication associated DNA glycosylase.[25] Furthermore, HCMV DNA polymerase (UL54) also interacts with UL114, suggesting the possibility that UL54, UL114 and UL44 form a heterotrimeric DNA replication complex.[26]

Chromatin remodelling is required for many DNA transactions, including transcription, replication and repair. One of the chromatin remodelling factors, the SWI/SNF core factor SMARCB1, forms a complex with UL114 (HCMV UNG) and the processivity factor UL44 in replication foci.[27] Previously, a proteomic analysis of purified murine CMV virions identified seven cellular proteins, including H2A but no other nucleosomal proteins.[28] Furthermore, early during HSV-infection, nucleosomes are associated with promoter regions of several transcribed viral genes, but are not present after onset of replication and in newly replicated viral DNA. Thus, the association of nucleosomes with herpes DNA during infection is temporary.[29] The possible significance of the interaction between

UL114/UL44 and SMARCB1 during replication remains unknown, but from the collective data, it appears likely that SMARCB1 is required in early stages of replication.

5.1.3. *Structure-function Relationships for Herpesvirus UNG*

The overall structure of HSV-1 UNG and its active site pocket[30] are very similar to the corresponding structures of human UNG[31,32] and *E. coli*.[33,34] However, there are significant structural differences in regions with insertions and low sequence homology and more subtle differences in the active site pockets in herpes UNG when compared with *E. coli*/human UNG. Overall, *E. coli* Ung is more similar to human UNG than to herpesvirus UNG. The differences between the active site of human UNG and herpesvirus UNG are largely restricted to side chain positions prior to substrate binding[34] (Fig. 5.1).

A direct comparison of human UNG and HSV-1 UNG indicated that the catalytic efficiency (k_{cat}/K_M) of the former is ~10–50-fold higher than that of HSV-1 UNG, due to higher turnover number (k_{cat}) and lower K_M of the human enzyme.[35] While this is likely to be correct, comparisons may be influenced by assay conditions as well as truncations of the human enzyme form used. Thus, it is known that the kinetic parameters of full length form of human nuclear UNG (hUNG2) is strongly influenced by the presence/absence of Mg^{2+} and AP-endonuclease, as well as type of DNA substrate [36]. However, the catalytic efficiencies reported for hUNG2[36] is generally much higher under different assay conditions than those of HSV-1.[35] Thus, both HSV-and HCMV have relatively low catalytic efficiency when compared to human UNG. Reconstitution of the BER pathway has been achieved using a combination of herpes UNG, herpes DNA polymerase, human AP-endonuclease (APE1) and human DNA ligase I or (preferably) human DNA ligase IIIα-XRCC1. This may be biologically important since APE1 and DNA ligase IIIα-XRCC1 are also expressed in non-proliferating cells. Furthermore, similar to DNA polymerase β, herpes DNA polymerase has apyrimidinic/apurinic and 5′-deoxyribose phosphate lyase activity.[37]

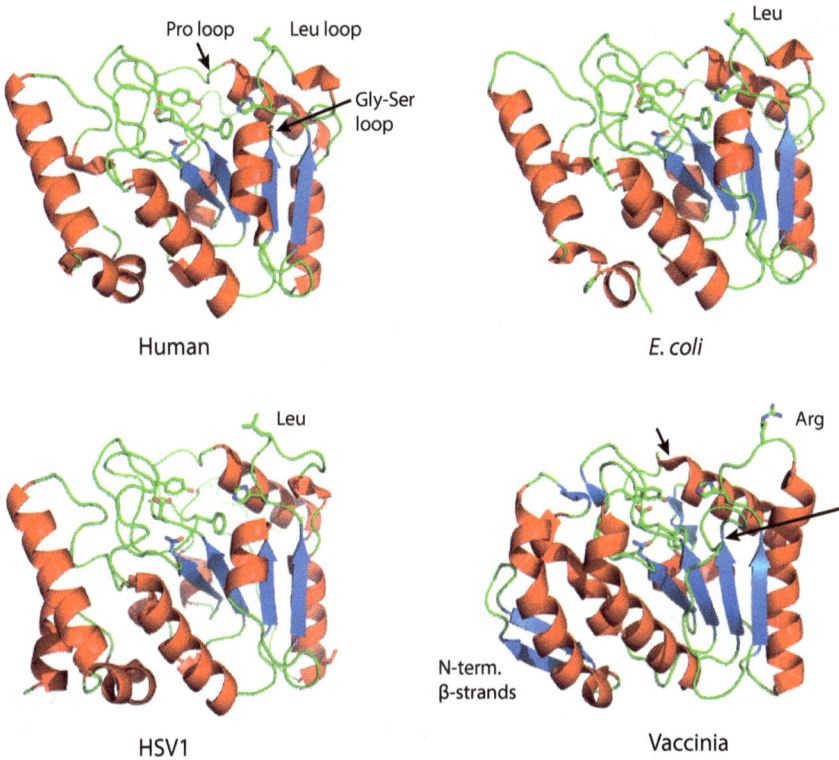

Fig. 5.1. Crystal structures of viral, bacterial and human UNG. The structures are visualized using PyMOL Molecular Graphics System using coordinates PDB ID: 1AKZ (human), 1UEG (*E. coli*), 1UDG (HSV1), and 5JX8 (vaccinia). Residue side chains important in substrate recognition (Asn, Tyr, Phe) and catalysis (Asp, His) are indicated. The proteins show high degree of structural conservation. However, the Pro-rich- and the Gly-Ser loops (indicated by arrows) are missing and the Leu in the DNA intercalating loop is not conserved in vaccinia virus UNG.

5.1.4. *Inhibitors of Herpesvirus UNG*

The strong requirement of herpesvirus UNG and dUTPase for neuro-virulence and reactivation from latency has triggered research on inhibitors of these enzymes as potential antiviral drugs. Inhibitors of herpesvirus DNA polymerase has for a long time been in clinical use to treat herpes infections. The nucleoside analogue acyclo-guanosine (acyclovir) is also widely used topically for various herpesvirus infections. Furthermore, infusion of acyclovir is approved for treatment of herpes simplex encephalitis. While clinical outcomes are still suboptimal, early treatment increases the chance of survival without serious neurological damage.[38] Acyclovir was the first effective antiviral drug to be registered and this landmark discovery was central when awarding the Nobel Prize in physiology or medicine to pharmacologist Gertrude B. Elion in 1988. Acyclovir, and the newer derivatives valacyclovir, ganciclovir and famciclovir are now used for a broad spectrum of herpes disease either topically or systemically. Immuno-compromised patients are more sensitive to herpesvirus infection and reactivation. In addition, they more frequently develop resistance to herpesvirus DNA polymerase inhibitors, such as acyclovir.[39] Inhibitors of herpesvirus UNG or dUTPase have so far not had the same kind of success. However, a number of uracil-analogs have been found to have some selectivity for herpes uracil-DNA glycosylase over the host enzyme.[40,41] Inhibitors of the 6-(4-alkylanilino)-uracil type that differentiate better between herpes and host UNG have been designed based on a computational rationale for selective inhibition.[42] However, we are not aware of clinical trials or *in vivo* studies in animals that have examined the potential of selective inhibitors of herpes UNG or dUTPase. One general and significant problem when treating herpes with DNA polymerase inhibitors is that treatment is effective only when started very early after infection or reactivation. Since herpes UNG and herpes dUTPase, like herpesvirus DNA polymerase, are among early expressed proteins, potential drugs against these targets would most likely also be effective only after early treatment.

5.2. Poxvirus Uracil-DNA Glycosylase and dUTPase

Poxviridae are a large family of viruses that infect vertebrates. They are enveloped viruses with large double stranded DNA genomes (130–230 kbp) that encode a large number of proteins, including DNA replication and DNA repair proteins. Two human poxviruses are known; variola virus causing the deadly smallpox (now eradicated) and molluscum contagiosum virus (MCV) causing benign skin lesions in healthy children, but serious disease in immunocompromised individuals. In addition, some other poxviruses having other species as their natural host can also be transmitted to humans, e.g. vaccinia virus (cowpox) and monkey poxvirus. Vaccinia virus infection in humans is mild and usually asymptomatic, and is the prototype poxvirus for molecular biology studies. It also, of course, reached fame because vaccinia virus infection protects humans against smallpox. This discovery represents an important part of the history of medicine and also gave name to the process of vaccination, *vacca* being the Latin word for cow.

Poxviruses replicate in replication factories in the cytoplasm, producing long DNA concatemers that are resolved by a viral Holliday junction resolvase.[43] These replication factories accumulate virus proteins for transcription, replication and repair and are temporarily ER membrane-bound. The replication factories also hijack several host cell translation factors that may enhance virus production and suppress host cell functions.[44]

5.2.1. *Poxvirus UNG (D4) is Essential as Part of a DNA Replication Complex*

A poxvirus gene encoding uracil-DNA glycosylase of the UNG-family was first identified in Shope fibroma virus[45] and shown to be essential for virus viability.[46] The D4R ORF encodes poxvirus UNG in Shope fibroma virus and several other family members, including vaccinia virus. The amino acid sequence of the vaccinia UNG (218 amino acids) shows only ~20% identity with human UNG and there are also significant differences in the secondary and tertiary structures.[47,48] Poxvirus UNG is thus a more distant member of the family than herpesvirus UNG.

Using a temperature-sensitive mutant in the protein D4 (named *ts*4149), vaccinia virus UNG was shown to be essential for virus replication.[49] The mutation in *ts*4149 was later identified as Gly179Arg. Mutagenesis back to Gly179 restored catalytic activity and viral replication. Furthermore, mutagenesis of three different active site pocket residues produced catalytically inactive UNG proteins that could still bind DNA, but not support replication, indicating that the catalytic activity of vaccinia virus UNG was required for virus proliferation and viability.[50] Successful complementation of the *ts*4149 mutant by engineered cell lines expressing wild-type vaccinia virus UNG demonstrated that viral UNG acts in *trans* and that glycosylase activity is required.[51] However, it was later reported that the role of viral UNG in virus DNA replication is independent of glycosylase activity, although virulence was decreased *in vivo* in mice by active site mutations.[52] These in part contradicting results might be due to different methods used. Furthermore, the catalytic activity of poxvirus UNG is 2–3 orders of magnitude lower than that of human UNG, making measurement of residual activity in mutants challenging.[53]

5.2.2. *Structure-function Relationship of Vaccinia Virus UNG*

The overall core structure of vaccinia virus UNG is well conserved. However, vaccinia virus UNG is a dimer both in solution and in the crystal structure and the overall structure contains an extra β-strand at each terminus, when compared with human UNG.[47] There are differences in the number and organization of α-helices as well. In the active site region there are also some differences. In human UNG and *E. coli* Ung, motifs that are important for catalysis are highly conserved, including the catalytic water-activating loop and other active site residues, uracil-specificity residues, the Pro-rich loop, the Gly-Ser loop and Leu-intercalation loop. In contrast, in vaccinia virus UNG (D4) there are some differences in all these motifs when compared with the *E. coli* or human enzyme, and some motifs are even missing (Pro-rich loop and Gly-Ser loop) or very different (Leu-intercalation loop).[47] However, the structure of the D4-uracil complex demonstrates that the uracil-binding pocket is highly conserved.[54] The amino acid change in *ts*4149 (Gly179Arg) in the vaccinia virus UNG mutant discussed above was found to be located at the C-terminal tip of β-strand 9 outside of the active site. It was suggested that the positive side chain of Arg179 in the mutant would rotate into a very hydrophobic pocket (Tyr156, Ile177, Phe195 and Ile198) and cause structural destabilization as a basis for its temperature sensitivity.[47]

5.2.3. *Vaccinia UNG (D4) is a Processivity Factor for Vaccinia Virus DNA Polymerase (E9)*

The poxvirus DNA polymerase catalytic subunit has polymerase and 3′–5′ proofreading exonuclease activities, similar to some, but not all, Family B DNA polymerases.[55] The isolated catalytic subunit of poxvirus DNA polymerase (E9) carries out distributive DNA synthesis,[56] but gains processivity when in complex the viral adaptor protein A20 and vaccinia UNG (D4). This heterotrimer constitutes the vaccinia DNA polymerase holoenzyme.[57,58] The N-terminal amino acids of A20 interact with the C-terminal region of D4.[58] Furthermore, a conserved hexamer stretch at the C-terminal end of D4 is required both for interaction with A20 and for protein integrity.[59] Thus, D4 is a uracil-DNA glycosylase, a DNA polymerase processivity factor and a protein stabilizing factor. Interestingly, positively charged amino acids in D4 required for processivity can be mutated without affecting glycosylase activity and binding to A20, indicating that these functions can be separated.[60] D4 binds weakly to uracil-containing DNA and to the phosphate backbone of undamaged DNA, but binds strongly to abasic sites.

Furthermore, the mode of binding and the enzyme orientation on DNA has diverged from that of human UNG.[61] A low-resolution structure of the E9-A20-D4 complex was obtained in small-angle-X-scattering (SAXS) studies. These studies demonstrated a 1:1:1 stoichiometry of the subunits E9-A20-D4 of the holoenzyme and further indicated that A20 bridges E9 and D4 without contacting DNA, whereas E9 and D4 both contact DNA (Fig. 5.2). The distance between catalytic site of E9 and the DNA-binding site of D4 is approximately 150 Å, which corresponds to 50–60 base pairs.[55,62]

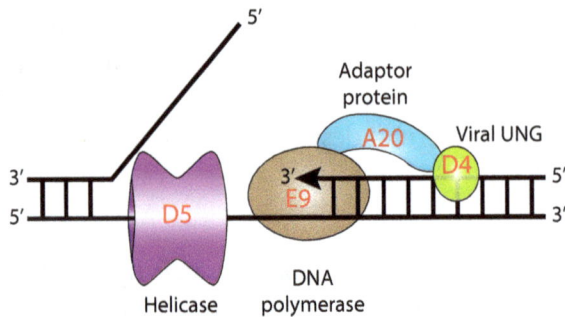

Fig. 5.2.   Model depicting a potential role for poxvirus UNG (D4) as part of a viral processivity complex during viral replication.

5.3. Retrovirus HIV-1 Incorporates UNG2 and Other Cellular Proteins in the Virus Particle

Several enveloped viruses incorporate host cell proteins into their virions. Different viruses incorporate different cellular proteins, indicating that it is not a random type of incorporation. Rather, incorporation of host proteins is a complex process that viruses use different host cell pathways to accomplish, such as the multivesicular body (MVB) pathway. The role of hijacked proteins is frequently not known. Human immunodeficiency virus type 1 (HIV-1) has been shown to incorporate a large number of host cell proteins, including several heat shock proteins, APOBEC3G, APOBEC3F and UNG (reviewed in[63]). The APOBEC proteins deaminate cytosines to uracil and may act as restriction factors in an innate immune response. The role of incorporated UNG remains controversial and has been reported to both prevent and promote viral infection, or to have no effect in pathogenesis. This will be discussed below, largely using HIV-1 as a basis. Some background information is required for this discussion.

5.3.1. *The Life Cycle of HIV-1*

HIV-1 is a retrovirus carrying a relatively small RNA genome of ~10 kb. The core of HIV-1 contains viral capsid proteins, reverse transcriptase (RT), integrase, Vpr and, as mentioned, several cellular proteins. RT is a low fidelity polymerase lacking a 3′–5′ proofreading exonuclease. It can synthesize DNA using either RNA or DNA as template. RT also has ribonuclease H (RNase H) activity that degrades the viral RNA from the DNA-RNA hybrid, generating a specific polypurine tract (PPT) RNA primer used for (+) strand DNA synthesis. The selection of the correct RNA primer is aided by a chaperone activity of HIV-1 nucleocapsid (NC) protein.[64] Vpr is a multifunctional protein perhaps best known as a factor inducing cell cycle arrest at the G2/M transition and depletion of CD4+ T cells, but it also has roles in modulating DNA repair, transcription, apoptosis and other processes.[65] In the life cycle of HIV-1, envelope glycoproteins first bind to receptors on the target cells (CD4+ T cells, macrophages and dendritic cells). After fusion of viral and cellular membranes, the viral core is released into the cytoplasm. The viral RNA genome is then reverse transcribed by RT to (−) strand DNA, which is used as template to produce (+) strand DNA.[66,67] HIV-1 double-stranded DNA forms a preintegration complex (PIC) together with viral integrase, capsid proteins and some host cell proteins. Since HIV-1 target cells do not proliferate, the large PIC enters nuclei via the nuclear pore channel complex in an energy-dependent process. Integration of viral DNA into cellular DNA requires viral integrase that binds to the LTR (long terminal repeat) ends of HIV double-stranded DNA. In CD4+ T cells, the viral genome enters a latency status where it can be transcriptionally silenced and thus a difficult target to reach for drugs.[66] Reactivation of the virus can make eradication by drugs easier, but this remains difficult due to multiple barriers.[68]

5.3.2. *Some Non-primate Lentiviruses have dUTPase Genes, Whereas Primate Lentiviruses may Package Cellular UNG2 in Virions*

The lentivirus equine infectious anaemia virus (EIAV) is an immunosuppressive virus that encodes its own dUTPase. Mutant viruses lacking dUTPase were found to have reduced capacity for integration of the virus and expression of viral genes, while at the same time genomic uracil accumulated due to incorporation of dUTP into DNA. Thus, expression of dUTPase by EIAV protects this virus from degradation and enhances virus yield and infectivity.[69] A related role of viral dUTPase was reported for the goat lentivirus caprine arthritis-encephalitis virus (CAEV). Furthermore, CAEV-mutants defective in dUTPase accumulated G to A transition mutations.[70] Primate lentiviruses do not have a gene for dUTPase. However, UNG incorporated in the virus particle and released upon infection could potentially have a similar role. Early work in this area detected UNG as a protein that interacts directly with the viral accessory protein Vpr, using a two-hybrid screen and co-immunoprecipitation as methods. Unlike several other UNG-interacting proteins, the interaction with Vpr did not require the N-terminal extensions of nuclear UNG2, suggesting that binding was via the catalytic domain of the enzyme. The UNG-Vpr complex was reported to be active, suggesting that it could serve a role in removing uracil from viral replication intermediates, as an alternative to the role of viral dUTPase in non-primate lentiviruses in *preventing* uracil incorporation.[71] In one study, the incorporation of UNG into HIV-1 was observed in Vpr-deficient virions, but required HIV-1 encoded integrase with which UNG interacts directly.[72] However, another study concluded that the major route for incorporation of UNG2 into HIV-1 virions is through interaction with Vpr, but did not exclude integrase-dependent incorporation as an alternative.[73] Although integrases from SIV and HIV-2 also interact weakly with UNG, incorporation of UNG2 into the virion apparently only occurs in HIV-1.[74] HIV-1 integrase variants carrying amino acid substitutions L172A or K173A failed to incorporate UNG2 in the virus particle. Furthermore, lysates of wild-type virions containing UNG2 was shown to process DNA oligonucleotides with a mismatched U:G at the primer end, whereas no such processing was observed in lysates of UNG2-deficient virions.[75] It should be noted that this processing is not representative of complete base excision repair.

5.3.3. *Does UNG2 Help HIV-1 Dissemination or Defend the Host Cell?*

There are two major mechanisms by which UNG2 incorporated in the HIV-1 virion could theoretically affect virus infectivity and function. The simplest idea is that UNG2 contributes to the repair of uracil in viral DNA intermediates, thus reducing the mutational load and enhancing infectivity of the virus. The other side of the coin would be that high levels of uracil in viral DNA intermediates, due to high dUTP levels in the cell, or deamination of cytosines to uracil by APOBEC-proteins, would lead to excessive strand breaks and viral DNA degradation initiated by UNG2. Both aspects have been explored, but a consensus on the outcome has not been reached. Most likely this may be explained by different properties of various targets, methods used as well as parameters tested. There is, however, a consensus opinion that uracilation of HIV-1 DNA due to deamination by APOBEC3F and/or APOBEC3G is part of an innate immune response to HIV-1 infection. However, the major source of HIV-1 uracilation in monocytes and monocyte-derived macrophages appears to be by dUTP incorporation. These cells have undetectable dUTPase and low UNG expression levels.[76]

Due to the low fidelity of RT, the mutation frequency during HIV-1 replication is very high. In addition, HIV-1 replication *in vivo* is highly productive, thus contributing to the high mutation frequency.[77] Uracil in HIV-1 DNA could be one source contributing to mutagenesis. Support for this hypothesis was obtained in studies demonstrating 3–4-fold increase in mutation frequencies in HIV-1 or HIV-1 DNA vectors when Vpr is absent or unable to interact with UNG2.[73,78] Furthermore, replication of HIV-1 in non-proliferating primary monocyte-derived macrophages that lack UNG2 resulted in 18-fold increase in mutation frequencies.[78] In general agreement with this, HIV-1-associated UNG2 was reported to be essential to the viral life cycle by controlling dUTP misincorporation.[79] In addition, it was reported that UNG2 and APE1 cooperate to degrade viral DNA and that APE1 may also be incorporated in the virion.[80] Other studies demonstrated that incorporation of UNG2 into virions was required for full infectivity and that this effect required interaction with the p32 subunit of replication protein A (RPA). The infectivity required interaction of RPA with UNG2, rather than the enzymatic activity of UNG2.[81,82] Interestingly; dUTP-incorporation in HIV-1 DNA in a cell free system had no effect on (−) strand synthesis, but significantly reduced the yield of (+) strand synthesis by reducing the specificity for initiation at the correct polypurine tract primer generated by the RNase H activity of RT.[83] It was later demonstrated that the fidelity of (+) strand priming requires the chaperone activity of HIV-1 nucleocapsid protein,[64] likely explaining the reduced yield of (+) strand synthesis in the cell free system lacking this chaperone. Also, UNG2 was reported to initiate degradation of HIV-1 cDNA containing misincorporated dUTP, thus preventing viral integration.[84]

Although several studies indicate an important role of UNG2 for the infectivity of HIV-1, the picture is far from clear. Thus, one study concluded that UNG2 does not have an effect on HIV-1 infectivity in a number of cell lines, as well as primary macrophages. Furthermore, the loss of infectivity caused by APOBEC3G was unaffected by the presence of UNG2, indicating that UNG2 is not required for innate responses against HIV-1.[85,86] The latter result was supported by the observation that although APOBEC3G reduced correct (+) strand synthesis and integration, this effect did not involve degradation of viral DNA by UNG2.[87] Importantly, several studies have demonstrated that UNG2 levels are very low in cells infected by HIV-1, due to ubiquitinylation and proteasomal degradation of UNG2, as well as SMUG1. This required Vpr in a lysate that also contained E3 ubiquitin ligase components CUL1 and CUL4.[88] More importantly, damage-specific DNA-binding protein 1 (DDB1), also known to be component of E3 ubiquitin ligase, interacts directly with Vpr and mediates UNG2/SMUG1 degradation.[89] The E3 ligase complex that UNG2 is associated with is composed of cullin 4A (CUL4A), RING H2 finger protein (RBX1), DDB1-and CUL4-associated factor 1 (DCAF1) and Vpr, the latter strongly facilitating ubiquitinylation and degradation of UNG2. In this complex Vpr mediates interaction between UNG2 and DCAF1.[90]

The crystal structure of the DDB1-DCAF1-VPR-UNG2 has now revealed in more detail how Vpr interacts to mediate degradation of UNG2. Briefly, Vpr forms a bridge between UNG2 and DCAF1, whereas DCAF1 also interacts directly with DDB1.[91] Superposition of this DDB1 structure with previously solved structures for DDB1-CUL4-RBX1 made possible the presentation of a likely molecular

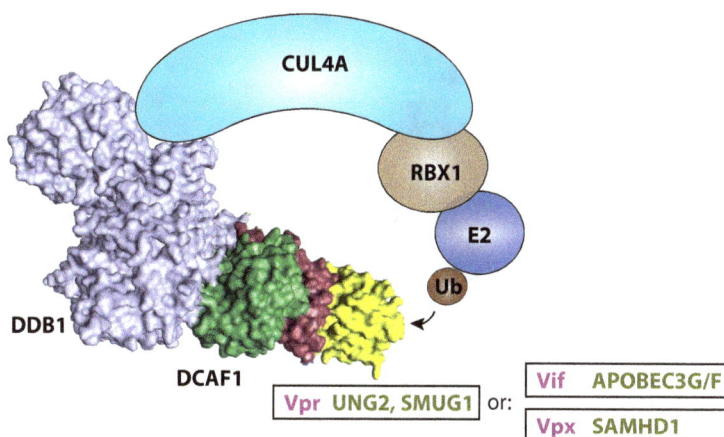

Fig. 5.3. HIV-1 Vpr mediates degradation of UNG2 by attachment to the CRL4-DCAF1 E3 ubiquitin ligase complex and subsequent ubiquitinylation. The DDB1-DCAF1-Vpr-UNG structure is from PDB ID: 5JK7. Notably, other viral proteins may mediate attachment of other target proteins for degradation to a similar E3 ligase complex. Thus, Vif and Vpx mediate attachment and degradation of APOBEC3G/F and SAMHD1, respectively, whereas Vpr may also mediate attachment of SMUG1.

model for the structure of the whole CRL4-DCAF1 E3 ubiquitin ligase complex.[91] A simplified cartoon-presentation of this model shows how the juxtaposition of UNG2 and E2 makes UNG2 available for ubiquitinylation (Fig. 5.3).

The crystal structure of Vpr forms a three-helix bundle (Fig. 5.4A, in red). Structural motifs in Vpr comprise the N-terminal random coil tail, a hydrophobic cleft formed by helices α1, α2 and the first turn of α3, the loop connecting α2 and α3 (insert loop) and the C-terminal part of helix α3. The binding of Vpr to UNG2 is via the insert loop and residues in the hydrophobic cleft. Interestingly, the Vpr-UNG2 interaction is very similar to the interaction of UNG with DNA and Ugi (Fig. 5.4B) and like Ugi, Vpr inhibits the enzyme activity. In the UNG-DNA complex, Leu272 of UNG is inserted into DNA via the minor groove (Fig. 5.4C). Similarly, Leu272 penetrates into the hydrophobic cleft in the Vpr-UNG2 complex. The interaction of Vpr with DCAF is via the N-terminus and helix α3.[91]

(A)   (B)

(C)

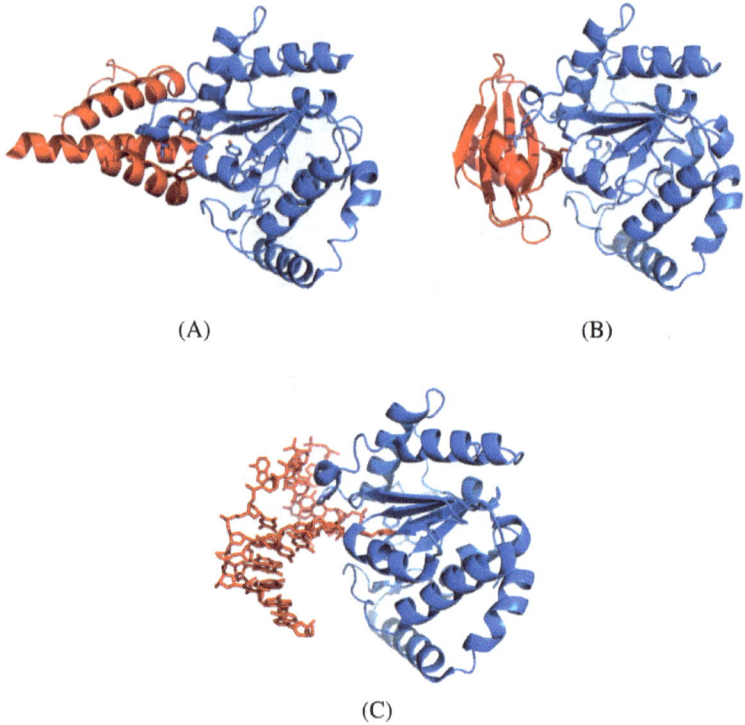

Fig. 5.4. Crystal structures of human UNG (blue) bound to HIV-1 Vpr (A), Ugi (B) and DNA (C). The structures are visualized using PyMOL Molecular Graphics System using coordinates PDB ID:5JK7, ID:1ugh and ID:1EMH, respectively. Vpr binds to the same area on the UNG surface as Ugi and double-stranded DNA.

Interestingly, the HIV-1 restriction factor SAMHD1 (SAM domain and HD domain-containing protein 1) binds to HIV-1 protein Vpx (a Vpr homolog) and mediates degradation of SAMHD1, most likely by a similar mechanism.[91] Furthermore, viral infectivity factor (Vif) acts by mediating ubiquitinylation and degradation of APOBEC3G, and apparently some other APOBEC3 proteins. Vif is also required for incorporation of APOBEC3G in virions. It should be appreciated that the E3 ligase complexes come in different variants. Thus, as an example, Vif is associated with CUL5 rather than CUL4.[92]

In conclusion, introduction of genomic uracil into HIV-1 DNA intermediates occurs by incorporation of dUTP during DNA synthesis and deamination of DNA-cytosine to uracil by APOBEC3G or APOBEC3F. Furthermore, UNG2, as well as APOBEC3G, are incorporated in HIV-1 virions and have been regarded as components in an innate immune response to the virus. Whereas this remains the current view for APOBEC3G/F, the possible role of UNG2 in initiating degradation of HIV-1 DNA remains controversial. As outlined above, some reports find that UNG2 prevents deleterious mutations in HIV-1 DNA, whereas others report that UNG2 contributes to DNA degradation, and some find neither protection nor degradation of viral DNA. There is, however, agreement that the HIV-1 accessory proteins Vpr, Vpx and Vif are involved in ubiquitinylation and subsequent proteasomal degradation of UNG2/SMUG1, SAMHD1 and APOBEC3G, respectively. Furthermore, SAMHD1 and APOBEC3G are accepted as important factors in the innate immune defence against HIV-1. Given phenotypic differences of HIV-1 target cells, as well as the complex effects of HIV accessory proteins on the cell cycle, transcription, apoptosis, DNA repair and protein stability, it is perhaps not surprising that the role of UNG2 during HIV-1 infection has proven difficult to define.

References

1. Deng, L., Qin, X., Krell, P., *et al.* Characterization and functional studies of fowl adenovirus 9 dUTPase. *Virology* **497**, 251–261 (2016).
2. Chung, S. H., Weiss, R. S., Frese, K. K., *et al.* Functionally distinct monomers and trimers produced by a viral oncoprotein. *Oncogene* **27**, 1412–1420 (2008).
3. Weiss, R. S., Lee, S. S., Prasad, B. V., Javier, R. T. Human adenovirus early region 4 open reading frame 1 genes encode growth-transforming proteins that may be distantly related to dUTP pyrophosphatase enzymes. *J Virol* **71**, 1857–1870 (1997).
4. Hizi, A., Herzig, E. dUTPase: the frequently overlooked enzyme encoded by many retroviruses. *Retrovirology* **12**, 70 (2015).
5. Kwon, E., Pathak, D., Chang, H. W., Kim, D. Y. Crystal structure of mimivirus uracil-DNA glycosylase. *PLoS One* **12**, e0182382 (2017).

6. Berges, B. K., Tanner, A. Modelling of human herpesvirus infections in humanized mice. *J Gen Virol* **95**, 2106–2117 (2014).

7. Cornaby, C., Tanner, A., Stutz, E. W., *et al.* Piracy on the molecular level: human herpesviruses manipulate cellular chemotaxis. *J Gen Virol* **97**, 543–560 (2016).

8. Grinde, B. Herpesviruses: latency and reactivation — viral strategies and host response. *J Oral Microbiol* **5** (2013).

9. Sorel, O., Dewals, B. G. MicroRNAs in large herpesvirus DNA genomes: recent advances. *Biomol Concepts* **7**, 229–239 (2016).

10. Weller, S. K., Coen, D. M. Herpes simplex viruses: mechanisms of DNA replication. *Cold Spring Harb Perspect Biol* **4**, a013011 (2012).

11. Caradonna, S. J., Cheng, Y. C. Induction of uracil-DNA glycosylase and dUTP nucleotidohydrolase activity in herpes simplex virus-infected human cells. *J Biol Chem* **256**, 9834–9837 (1981).

12. Caradonna, S. J., Adamkiewicz, D. M. Purification and properties of the deoxyuridine triphosphate nucleotidohydrolase enzyme derived from HeLa S3 cells. Comparison to a distinct dUTP nucleotidohydrolase induced in herpes simplex virus-infected HeLa S3 cells. *J Biol Chem* **259**, 5459–5464 (1984).

13. Barker, D. E., Roizman, B. Identification of three genes nonessential for growth in cell culture near the right terminus of the unique sequences of long component of herpes simplex virus 1. *Virology* **177**, 684–691 (1990).

14. Pyles, R. B., Sawtell, N. M., Thompson, R. L. Herpes simplex virus type 1 dUTPase mutants are attenuated for neurovirulence, neuroinvasiveness, and reactivation from latency. *J Virol* **66**, 6706–6713 (1992).

15. Pyles, R. B., Thompson, R. L. Evidence that the herpes simplex virus type 1 uracil DNA glycosylase is required for efficient viral replication and latency in the murine nervous system. *J Virol* **68**, 4963–4972 (1994).

16. Pyles, R. B., Thompson, R. L. Mutations in accessory DNA replicating functions alter the relative mutation frequency of herpes simplex virus type 1 strains in cultured murine cells. *J Virol* **68**, 4514–4524 (1994).

17. Worrad, D. M., Caradonna, S. Identification of the coding sequence for herpes simplex virus uracil-DNA glycosylase. *J Virol* **62**, 4774–4777 (1988).

18. Mullaney, J., Moss, H. W., McGeoch, D. J. Gene UL2 of herpes simplex virus type 1 encodes a uracil-DNA glycosylase. *J Gen Virol* **70** (**Pt 2**), 449–454 (1989).

19. Varshney, U., Hutcheon, T., van de Sande, J. H. Sequence analysis, expression, and conservation of Escherichia coli uracil DNA glycosylase and its gene (ung). *J Biol Chem* **263**, 7776–7784 (1988).

20. Plosa, E. J., Esbenshade, J. C., Fuller, M. P., Weitkamp, J. H. Cytomegalovirus infection. *Pediatr Rev* **33**, 156–163; quiz 163 (2012).

21. Prichard, M. N., Duke, G. M., Mocarski, E. S. Human cytomegalovirus uracil DNA glycosylase is required for the normal temporal regulation of both DNA synthesis and viral replication. *J Virol* **70**, 3018–3025 (1996).

22. Penfold, M. E., Mocarski, E. S. Formation of cytomegalovirus DNA replication compartments defined by localization of viral proteins and DNA synthesis. *Virology* **239**, 46–61 (1997).

23. Courcelle, C. T., Courcelle, J., Prichard, M. N., Mocarski, E. S. Requirement for uracil-DNA glycosylase during the transition to late-phase cytomegalovirus DNA replication. *J Virol* **75**, 7592–7601 (2001).

24. Prichard, M. N., Lawlor, H., Duke, G. M., *et al.* Human cytomegalovirus uracil DNA glycosylase associates with ppUL44 and accelerates the accumulation of viral DNA. *Virol J* **2**, 55 (2005).

25. Ranneberg-Nilsen, T., Dale, H. A., Luna, L., *et al.* Characterization of human cytomegalovirus uracil DNA glycosylase (UL114) and its interaction with poly-merase processivity factor (UL44). *J Mol Biol* **381**, 276–288 (2008).

26. Strang, B. L., Coen, D. M. Interaction of the human cytomegalovirus uracil DNA glycosylase UL114 with the viral DNA polymerase catalytic subunit UL54. *J Gen Virol* **91**, 2029–2033 (2010).

27. Ranneberg-Nilsen, T., Rollag, H., Slettebakk, R., *et al.* The chromatin remodel-ing factor SMARCB1 forms a complex with human cytomegalovirus proteins UL114 and UL44. *PLoS One* **7**, e34119 (2012).

28. Kattenhorn, L. M., Mills, R., Wagner, M., *et al.* Identification of proteins associated with murine cytomegalovirus virions. *J Virol* **78**, 11187–11197 (2004).

29. Oh, J., Fraser, N. W. Temporal association of the herpes simplex virus genome with histone proteins during a lytic infection. *J Virol* **82**, 3530–3537 (2008).

30. Savva, R., McAuley-Hecht, K., Brown, T., Pearl, L. The structural basis of specific base-excision repair by uracil-DNA glycosylase. *Nature* **373**, 487–493 (1995).

31. Mol, C. D., Arvai, A. S., Slupphaug, G., *et al.* Crystal structure and mutational analysis of human uracil-DNA glycosylase: structural basis for specificity and catalysis. *Cell* **80**, 869–878 (1995).

32. Slupphaug, G., Mol, C. D., Kavli, B., *et al.* A nucleotide-flipping mechanism from the structure of human uracil-DNA glycosylase bound to DNA. *Nature* **384**, 87–92 (1996).

33. Ravishankar, R., Bidya Sagar, M., Roy, S., *et al.* X-ray analysis of a complex of Escherichia coli uracil DNA glycosylase (EcUDG) with a proteinaceous inhibitor.

The structure elucidation of a prokaryotic UDG. *Nucleic Acids Res* **26**, 4880–4887 (1998).

34. Xiao, G., Tordova, M., Jagadeesh, J., *et al.* Crystal structure of Escherichia coli uracil DNA glycosylase and its complexes with uracil and glycerol: structure and glycosylase mechanism revisited. *Proteins* **35**, 13–24 (1999).

35. Krusong, K., Carpenter, E. P., Bellamy, S. R., *et al.* A comparative study of uracil-DNA glycosylases from human and herpes simplex virus type 1. *J Biol Chem* **281**, 4983–4992 (2006).

36. Kavli, B., Sundheim, O., Akbari, M., *et al.* hUNG2 is the major repair enzyme for removal of uracil from U:A matches, U:G mismatches, and U in single-stranded DNA, with hSMUG1 as a broad specificity backup. *J Biol Chem* **277**, 39926–39936 (2002).

37. Bogani, F., Chua, C. N., Boehmer, P. E. Reconstitution of uracil DNA glycosylase-initiated base excision repair in herpes simplex virus-1. *J Biol Chem* **284**, 16784–16790 (2009).

38. Gnann, J. W., Jr., Whitley, R. J. Herpes Simplex Encephalitis: an Update. *Curr Infect Dis Rep* **19**, 13 (2017).

39. Field, H. J., Vere Hodge, R. A. Recent developments in anti-herpesvirus drugs. *Br Med Bull* **106**, 213–249 (2013).

40. Focher, F., Verri, A., Spadari, S., *et al.* Herpes simplex virus type 1 uracil-DNA glycosylase: isolation and selective inhibition by novel uracil derivatives. *Biochem J* **292 (Pt 3)**, 883–889 (1993).

41. Sekino, Y., Bruner, S. D., Verdine, G. L. Selective inhibition of herpes simplex virus type-1 uracil-DNA glycosylase by designed substrate analogs. *J Biol Chem* **275**, 36506–36508 (2000).

42. Hendricks, U., Crous, W., Naidoo, K. J. Computational rationale for the selective inhibition of the herpes simplex virus type 1 uracil-DNA glycosylase enzyme. *J Chem Inf Model* **54**, 3362–3372 (2014).

43. Moss, B. Poxvirus DNA replication. *Cold Spring Harb Perspect Biol* **5**, (2013).

44. Katsafanas, G. C., Moss, B. Colocalization of transcription and translation within cytoplasmic poxvirus factories coordinates viral expression and subjugates host functions. *Cell Host Microbe* **2**, 221–228 (2007).

45. Upton, C., Stuart, D. T., McFadden, G. Identification of a poxvirus gene encoding a uracil DNA glycosylase. *Proc Natl Acad Sci U S A* **90**, 4518–4522 (1993).

46. Stuart, D. T., Upton, C., Higman, M. A., *et al.* A poxvirus-encoded uracil DNA glycosylase is essential for virus viability. *J Virol* **67**, 2503–2512 (1993).

47. Schormann, N., Grigorian, A., Samal, A., *et al.* Crystal structure of vaccinia virus uracil-DNA glycosylase reveals dimeric assembly. *BMC Struct Biol* **7**, 45 (2007).

48. Schormann, N., Ricciardi, R., Chattopadhyay, D. Uracil-DNA glycosylases-structural and functional perspectives on an essential family of DNA repair enzymes. *Protein Sci* **23**, 1667–1685 (2014).

49. Millns, A. K., Carpenter, M. S., DeLange, A. M. The vaccinia virus-encoded uracil DNA glycosylase has an essential role in viral DNA replication. *Virology* **198**, 504–513 (1994).

50. Ellison, K. S., Peng, W., McFadden, G. Mutations in active-site residues of the uracil-DNA glycosylase encoded by vaccinia virus are incompatible with virus viability. *J Virol* **70**, 7965–7973 (1996).

51. Holzer, G. W., Falkner, F. G. Construction of a vaccinia virus deficient in the essential DNA repair enzyme uracil DNA glycosylase by a complementing cell line. *J Virol* **71**, 4997–5002 (1997).

52. De Silva, F. S., Moss, B. Vaccinia virus uracil DNA glycosylase has an essential role in DNA synthesis that is independent of its glycosylase activity: catalytic site mutations reduce virulence but not virus replication in cultured cells. *J Virol* **77**, 159–166 (2003).

53. Scaramozzino, N., Sanz, G., Crance, J. M., *et al.* Characterisation of the substrate specificity of homogeneous vaccinia virus uracil-DNA glycosylase. *Nucleic Acids Res* **31**, 4950–4957 (2003).

54. Schormann, N., Banerjee, S., Ricciardi, R., Chattopadhyay, D. Structure of the uracil complex of Vaccinia virus uracil DNA glycosylase. *Acta Crystallogr Sect F Struct Biol Cryst Commun* **69**, 1328–1334 (2013).

55. Czarnecki, M. W., Traktman, P. The vaccinia virus DNA polymerase and its processivity factor. *Virus Res* (2017).

56. McDonald, W. F., Traktman, P. Vaccinia virus DNA polymerase. *In vitro* analysis of parameters affecting processivity. *J Biol Chem* **269**, 31190–31197 (1994).

57. Klemperer, N., McDonald, W., Boyle, K., *et al.* The A20R protein is a stoichiometric component of the processive form of vaccinia virus DNA polymerase. *J Virol* **75**, 12298–12307 (2001).

58. Ishii, K., Moss, B. Mapping interaction sites of the A20R protein component of the vaccinia virus DNA replication complex. *Virology* **303**, 232–239 (2002).

59. Nuth, M., Guan, H., Ricciardi, R. P. A Conserved Tripeptide Sequence at the C Terminus of the Poxvirus DNA Processivity Factor D4 Is Essential for Protein Integrity and Function. *J Biol Chem* **291**, 27087–27097 (2016).

60. Druck Shudofsky, A. M., Silverman, J. E., Chattopadhyay, D., Ricciardi, R. P. Vaccinia virus D4 mutants defective in processive DNA synthesis retain binding to A20 and DNA. *J Virol* **84**, 12325–12335 (2010).

61. Burmeister, W. P., Tarbouriech, N., Fender, P., *et al.* Crystal Structure of the Vaccinia Virus Uracil-DNA Glycosylase in Complex with DNA. *J Biol Chem* **290**, 17923–17934 (2015).

62. Sele, C., Gabel, F., Gutsche, I., *et al.* Low-resolution structure of vaccinia virus DNA replication machinery. *J Virol* **87**, 1679–1689 (2013).

63. Cantin, R., Methot, S., Tremblay, M. J. Plunder and stowaways: incorporation of cellular proteins by enveloped viruses. *J Virol* **79**, 6577–6587 (2005).

64. Post, K., Kankia, B., Gopalakrishnan, S., *et al.* Fidelity of plus-strand priming requires the nucleic acid chaperone activity of HIV-1 nucleocapsid protein. *Nucleic Acids Res* **37**, 1755–1766 (2009).

65. Soares, R., Rocha, G., Melico-Silvestre, A., Goncalves, T. HIV1-viral protein R (Vpr) mutations: associated phenotypes and relevance for clinical pathologies. *Rev Med Virol* **26**, 314–329 (2016).

66. Lusic, M., Siliciano, R. F. Nuclear landscape of HIV-1 infection and integration. *Nat Rev Microbiol* **15**, 69–82 (2017).

67. Menendez-Arias, L., Sebastian-Martin, A., Alvarez, M. Viral reverse transcriptases. *Virus Res* (2016).

68. Shan, L., Siliciano, R. F. From reactivation of latent HIV-1 to elimination of the latent reservoir: the presence of multiple barriers to viral eradication. *Bioessays* **35**, 544–552 (2013).

69. Steagall, W. K., Robek, M. D., Perry, S. T., Fuller, F. J., Payne, S. L. Incorporation of uracil into viral DNA correlates with reduced replication of EIAV in macrophages. *Virology* **210**, 302–313 (1995).

70. Turelli, P., Guiguen, F., Mornex, J. F., Vigne, R., Querat, G. dUTPase-minus caprine arthritis-encephalitis virus is attenuated for pathogenesis and accumulates G-to-A substitutions. *J Virol* **71**, 4522–4530 (1997).

71. Bouhamdan, M., Benichou, S., Rey, F., *et al.* Human immunodeficiency virus type 1 Vpr protein binds to the uracil DNA glycosylase DNA repair enzyme. *J Virol* **70**, 697–704 (1996).

72. Willetts, K. E., Rey, F., Agostini, I., *et al.* DNA repair enzyme uracil DNA glycosylase is specifically incorporated into human immunodeficiency virus type 1 viral particles through a Vpr-independent mechanism. *J Virol* **73**, 1682–1688 (1999).

73. Mansky, L. M., Preveral, S., Selig, L., Benarous, R., Benichou, S. The interaction of vpr with uracil DNA glycosylase modulates the human immunodeficiency virus type 1 *In vivo* mutation rate. *J Virol* **74**, 7039–7047 (2000).

74. Priet, S., Navarro, J. M., Gros, N., Querat, G., Sire, J. Differential incorporation of uracil DNA glycosylase UNG2 into HIV-1, HIV-2, and SIV(MAC) viral particles. *Virology* **307**, 283–289 (2003).

75. Priet, S., Navarro, J. M., Gros, N., Querat, G., Sire, J. Functional role of HIV-1 virion-associated uracil DNA glycosylase 2 in the correction of G:U mispairs to G:C pairs. *J Biol Chem* **278**, 4566–4571 (2003).

76. Hansen, E. C., Ransom, M., Hesselberth, J. R., *et al.* Diverse fates of uracilated HIV-1 DNA during infection of myeloid lineage cells. *Elife* **5** (2016).

77. Ho, D. D., Neumann, A. U., Perelson, A. S., *et al.* Rapid turnover of plasma virions and CD4 lymphocytes in HIV-1 infection. *Nature* **373**, 123–126 (1995).

78. Chen, R., Le Rouzic, E., Kearney, J. A., *et al.* Vpr-mediated incorporation of UNG2 into HIV-1 particles is required to modulate the virus mutation rate and for replication in macrophages. *J Biol Chem* **279**, 28419–28425 (2004).

79. Priet, S., Gros, N., Navarro, J. M., *et al.* HIV-1-associated uracil DNA glycosylase activity controls dUTP misincorporation in viral DNA and is essential to the HIV-1 life cycle. *Mol Cell* **17**, 479–490 (2005).

80. Yang, B., Chen, K., Zhang, C., Huang, S., Zhang, H. Virion-associated uracil DNA glycosylase-2 and apurinic/apyrimidinic endonuclease are involved in the degradation of APOBEC3G-edited nascent HIV-1 DNA. *J Biol Chem* **282**, 11667–11675 (2007).

81. Guenzel, C. A., Herate, C., Le Rouzic, E., *et al.* Recruitment of the nuclear form of uracil DNA glycosylase into virus particles participates in the full infectivity of HIV-1. *J Virol* **86**, 2533–2544 (2012).

82. Herate, C., Vigne, C., Guenzel, C. A., *et al.* Uracil DNA glycosylase interacts with the p32 subunit of the replication protein A complex to modulate HIV-1 reverse transcription for optimal virus dissemination. *Retrovirology* **13**, 26 (2016).

83. Klarmann, G. J., Chen, X., North, T. W., Preston, B. D. Incorporation of uracil into minus strand DNA affects the specificity of plus strand synthesis initiation during lentiviral reverse transcription. *J Biol Chem* **278**, 7902–7909 (2003).

84. Weil, A. F., Ghosh, D., Zhou, Y., *et al.* Uracil DNA glycosylase initiates degradation of HIV-1 cDNA containing misincorporated dUTP and prevents viral integration. *Proc Natl Acad Sci U S A* **110**, E448–457 (2013).

85. Kaiser, S. M., Emerman, M. Uracil DNA glycosylase is dispensable for human immunodeficiency virus type 1 replication and does not contribute to the antiviral effects of the cytidine deaminase Apobec3G. *J Virol* **80**, 875–882 (2006).

86. Schumacher, A. J., Hache, G., Macduff, D. A., *et al.* The DNA deaminase activity of human APOBEC3G is required for Ty1, MusD, and human immunodeficiency virus type 1 restriction. *J Virol* **82**, 2652–2660 (2008).

87. Mbisa, J. L., Barr, R., Thomas, J. A., *et al.* Human immunodeficiency virus type 1 cDNAs produced in the presence of APOBEC3G exhibit defects in plus-strand DNA transfer and integration. *J Virol* **81**, 7099–7110 (2007).

88. Schrofelbauer, B., Yu, Q., Zeitlin, S. G., Landau, N. R. Human immunodeficiency virus type 1 Vpr induces the degradation of the UNG and SMUG uracil-DNA glycosylases. *J Virol* **79**, 10978–10987 (2005).

89. Schrofelbauer, B., Hakata, Y., Landau, N. R. HIV-1 Vpr function is mediated by interaction with the damage-specific DNA-binding protein DDB1. *Proc Natl Acad Sci U S A* **104**, 4130–4135 (2007).

90. Ahn, J., Vu, T., Novince, Z., *et al.* HIV-1 Vpr loads uracil DNA glycosylase-2 onto DCAF1, a substrate recognition subunit of a cullin 4A-ring E3 ubiquitin ligase for proteasome-dependent degradation. *J Biol Chem* **285**, 37333–37341 (2010).

91. Wu, Y., Zhou, X., Barnes, C. O., *et al.* The DDB1-DCAF1-Vpr-UNG2 crystal structure reveals how HIV-1 Vpr steers human UNG2 toward destruction. *Nat Struct Mol Biol* **23**, 933–940 (2016).

92. Feng, Y., Baig, T. T., Love, R. P., Chelico, L. Suppression of APOBEC3-mediated restriction of HIV-1 by Vif. *Front Microbiol* **5**, 450 (2014).

Chapter 6

Genomic Uracil and Immunity

Bodil Kavli* and Geir Slupphaug

Identification of uracil in DNA as a key intermediate in both adaptive and innate immunity has made research on the introduction and fate of this non-canonical DNA base essential for understanding how we defend ourselves against invading pathogenic microorganisms. All cells have repair systems that remove DNA lesions to preserve genetic integrity. Occasionally, however, some lesions escape repair and result in changes in the DNA sequence. Together with selection of the fittest, this is the concept of evolution of all species. To combat pathogenic bacteria and virus (antigens), specific immune cells (B lymphocytes/B cells) have developed a sophisticated high-speed gene-targeted evolutionary process, termed somatic hypermutation (SHM). This extraordinary mechanism makes B cells capable of producing "tailor-made" high affinity antibodies (immunoglobulins) against any antigen. To initiate SHM, activated B-cells express an enzyme; activation-induced deaminase (AID) that introduces uracil in specific regions of the immunoglobulin (Ig) genes. These uracils are generally not repaired by error free BER, but are processed to generate point mutations or double strand breaks that modulate the antibody binding site or initiate Ig isotype switching, respectively. Cells expressing membrane-bound Igs (B-cell receptors) with increased

---

* bodil.kavli@ntnu.no

affinity for antigens are selected to proliferate and differentiate into antibody-secreting plasma cells and memory cells, which combat current infections and protect us against similar future pathogens.

Importantly, we have learned to exploit this adaptive immunological memory to improve human health by implementation of national and global vaccination programs. Vaccines are usually non-pathogenic antigenic material representing a specific virus or bacterium that is administrated into the body. This stimulates development of immunity to the pathogen by producing antibodies and cells with immunological memory and, in fact, genomic uracil is a key to develop this immunity. Vaccination has had a tremendous impact on human health and has eradicated or strongly reduced grave diseases such as smallpox and poliomyelitis and several million lives are saved each year due to various vaccination programs.

Uracil in DNA is also an intermediate in innate immunity. The APOBEC3 enzymes, which belong to the same deaminase protein family as AID, constitute a relatively newly discovered and unique defense mechanism against retrovirus (reviewed in Ref. 1). These enzymes act on cDNA after reverse transcription of the viral RNA genome, and restrict virus by introducing mutations and/or degradation of viral DNA.

6.1. Adaptive Immunity

The humoral arm of adaptive immunity is based on production of antibodies that can recognize the invading pathogen. However, independent of antigen stimulation, millions of B lymphocytes are generated in the bone marrow every day, and each developing B cell expresses a unique variant of the IgM B-cell receptor (BCR). BCRs and soluble antibodies consist of two identical immunoglobulin heavy chains (IgH) and two identical light chains of kappa (Igκ) or lambda (Igλ) type. Moreover, they have a variable antigen-binding site and a constant region that mediates the effector function (Fig. 6.1).

The human germ-line *Ig* genes (*IgH*, *Igκ*, *Igλ*) are located on different chromosomes and have a very unusual structure. The *IgH* gene

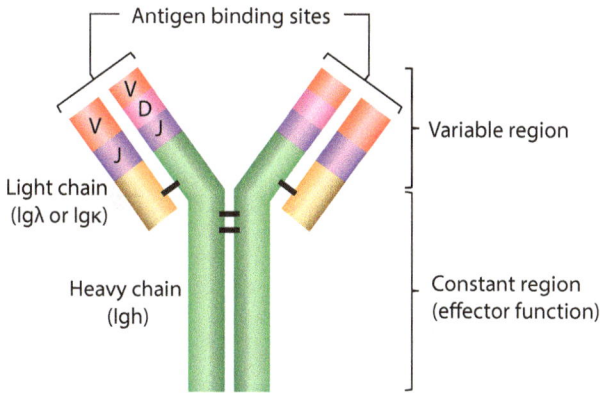

Fig. 6.1. Simplified structure of immunoglobulin antibody molecule

contains an array of ~40 V segments, 23 D segments and 6 functional J segments, in addition to the constant region exons that determines the Ig isotype (IgM, IgG, IgE and IgA). To generate functional antigen receptors during B cell development in the bone marrow, the variable domain exons are assembled by combining single gene segments to one coding VDJ exon.

This V(D)J recombination process is initiated by the recombination-activating gene proteins RAG1 and RAG2 (reviewed in[2]), a protein complex that introduces DNA double-strand breaks at specific sites between the gene segments. Single V, D and J segments are then joined together by the non-homologous end joining (NHEJ) pathway. Recombination also occurs in the light chain genes, Igκ and Igλ, but they consist only of V and J segment arrays. V(D)J recombination generates functional genes that encode the primary BCR repertoire on naïve B lymphocytes.

Mature naïve B cells are released from the bone marrow to the blood and migrate to secondary lymphoid organs (spleen and lymph nodes). Here they are activated by antigens, usually presented on the surface of antigen-presenting dendritic cells, and stimulated by T follicular helper cells (T$_{FH}$ cells) to form highly proliferative structures called germinal centers (Fig. 6.2). In the dark zone of these structures, two diversification mechanisms that alter the DNA sequence of the *Ig* genes occur: Generation of point mutations in the gene region

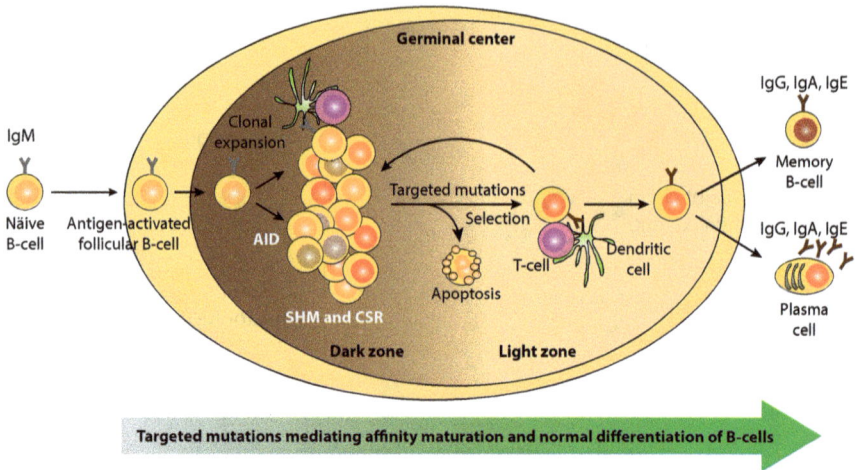

Fig. 6.2. Clonal expansion, affinity maturation (SHM) and isotype switching (CSR) occur within germinal centers in spleen and lymph node. Only B cells with increased affinity for antigen is selected to differentiate to antibody secreting plasma cells and memory B cells. T_{FH} cells and dendritic cells are involved in activation and selection of B-cell clones.

that code the antigen-binding site of the antibody (SHM), and switching of antibody isotype, typically from IgM to IgA, IgG or IgE, by class switch recombination (CSR). CSR involves recombination between two different *IgH* constant segments and results in a new biological effector function of the antibody without changing its antigen affinity. Cells with a functional BCR with increased affinity for antigen, and usually with new isotype, will survive (escape from apoptosis) and differentiate to plasma cells and memory B cells. Although the outcome of SHM and CSR is different, both processes are initiated by enzymatic introduction of uracil in specific regions of the *Ig* loci by the DNA deaminase AID (Fig. 6.3). Thus, AID-induced genomic uracil-dependent mutation- and recombination processes, together with selection, is the core mechanistic concept of adaptive immunity.

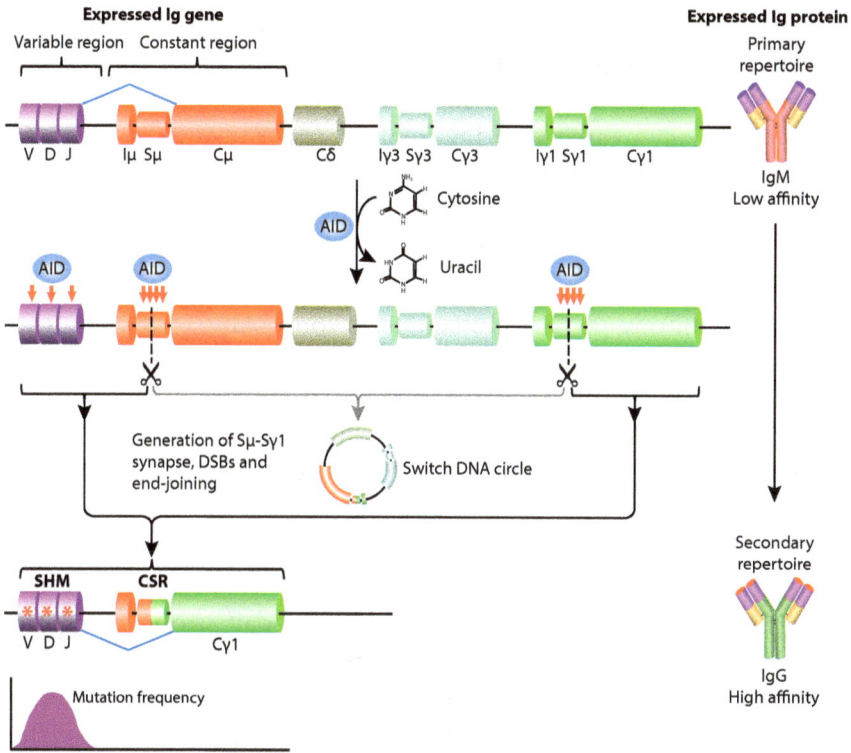

Fig. 6.3. AID initiates SHM and CSR by deaminating cytosine to uracil in Ig variable region and Ig switch regions, respectively. The figure illustrates the mutation and recombination events that occur in the Ig heavy chain locus in a cell that switches from expressing low affinity IgM to high affinity IgG antibody.

6.1.1. *Activation-induced Deaminase, AID*

AID was discovered in 1999 as a protein that was strongly induced in a stimulated B-cell lymphoma line and in primary mouse B-cells.[3] AID-deficient mice and human patients showed that AID is essential for both SHM and CSR.[4,5] Patients with inactivating mutations in the AID gene (*AICDA*) suffer from an autosomal recessive form of B-cell intrinsic Ig-CSR deficiency, previously termed hyper-IgM syndrome, with the antibody repertoire restricted to unmutated IgM antibodies (reviewed in Ref. 6). These patients are susceptible to infections caused by bacterial pathogens, especially those involving the respiratory and digestive tracts, and lymphoid hyperplasia is a hallmark. The disease is usually treated with IgG replacement therapy.

The ground-breaking discovery of AID made it possible to finally unravel the molecular mechanism of adaptive immunity. AID is found in all vertebrates including cartilaginous fish, which is the earliest vertebrate (~400 million years) to possess an adaptive immune system based on antibody genes.[7] Human and mouse AID is a small positively charged enzyme consisting of only 198 amino acids. The AID enzyme is member of a protein family of Zn-binding polynucleotide deaminases, the APOBECs. It was initially suggested to be an RNA deaminase due to its homology with APOBEC1,[3,4] which was the first member of this polynucleotide deaminase family to be discovered and characterized.[8] APOBEC1 deaminates apoliproprotein B mRNA at one specific cytosine (C_{6666}), resulting in a truncated isoform of the protein. However, the weight of evidence today strongly demonstrates that antibody diversification is triggered by direct DNA deamination by AID in the *Ig* loci, followed by processing of the resulting genomic uracil. Important for this conclusion is the finding that the uracil-DNA glycosylase activity of UNG is required for CSR and influences the SHM spectrum.[9-12] UNG would only play this role if uracil in DNA is an important key intermediate in the processes, and indeed AID-dependent introduction of uracil in Ig regions has been demonstrated by several methods.[13,14]

AID deaminates cytosine only in single-stranded DNA *in vitro* and in transcribed DNA *in vivo*, and prefers the sequence context WR$\underline{C}$ (W=A/T, R=purine). This is observed both in simple biochemical assays *in vitro* and *in vivo* as mutational hotspots in Ig variable and switch regions (reviewed in[15]).

The sequence preference of AID is directed by a specificity loop motif (Fig. 6.4), which is a conserved feature of the APOBEC family proteins. Switching this loop domain of AID with the specificity loop from other family members, changes the substrate preference accordingly.[16] The importance of this loop for sequence specificity was recently also demonstrated by the crystal structure AID[17] and the

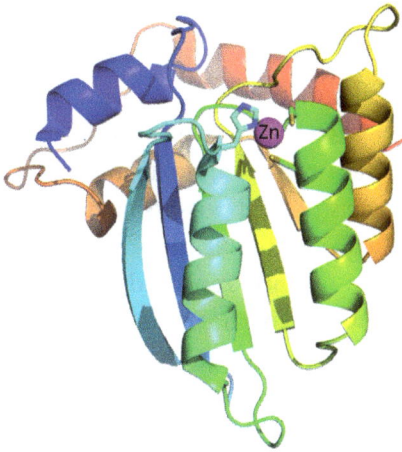

Fig. 6.4. Crystal structure of human AID. The figure is prepared with PyMOL Molecular Graphics System using coordinates for AIDv(Δ15) PDB ID:5JJ4.[17] The peptide chain is coloured from blue to red (rainbow), representing from N- to C-terminal direction. The Zn^{2+} ion (ball) and the conserved Zn-coordinating residue side chains (His56, Cys80, Cys87) are visualized. The DNA sequence specificity loop is shown in yellow.

structures of APOBEC3A and 3B bound to single-stranded DNA.[18] The DNA/enzyme complex structures demonstrate that the base at position −1 in the substrate fits into a hydrophobic pocket generated by the loop and that the specificity is provided by direct hydrogen bonds between the base and the protein. Compared to the APOBECs, AID has an extended specificity loop that explains the preference for a larger purine over pyrimidine residue 5′ prime to the target cytosine.

Although the primary targets for cytosine deamination by AID in normal B cells are variable and switch regions in the *Ig* loci, AID also deaminates other sites in the genome, including sites within several cancer-related genes.[19] This untargeted activity is linked to mutations and trans-locations at oncogenes in B-cell lymphoma (reviewed in Ref. 20 and covered in Chapter 7). In accordance with this, AID-expressing cancer cells accumulate genomic uracil and AID mutational signatures.[21,22]

In addition to its established role in adaptive immunity and mutagenesis, AID has been implicated in active DNA demethylation through its deaminase activity on 5-methylcytosine (5-mC) coupled with TDG induced BER.[23-25] *In vitro*, AID can deaminate 5-mC to T in ss DNA, although at a much lower efficiency than C to U deamination.[26,27] An alternative scenario to this pathway is that AID induces active DNA-demethylation by deamination of unmethylated cytosine, followed by UNG2 excision and possessive long-patch repair that will lead to removal of 5-mCs in the vicinity.[28]

Interestingly, it was recently reported that AID mutational signatures overlap with CpG methylation sites in many human cancers.[29] This indicates that unregulated AID activity may trigger tumorigenesis via deamination of both C and 5-mC. However, similarly to AIDs gene-specific function during Ig diversification, AID-mediated active DNA demethylation seems to occur only on selected pluripotency genes during cell differentiation (reviewed in[30]). Due to its mutagenic and likely also epigenetic activity, it is critical that AID is tightly controlled in the cell to achieve immunological diversity and normal cell differentiation without compromising genomic integrity.

6.1.2. *Regulation of AID*

Regulation of AID has mainly been studied in human cell lines and in mouse models. Cells use several mechanisms to tightly control AID expression and its biological activity. This is achieved through multilayered regulation of the AID gene (*AICDA* in humans and *Aicda* in mice*)*, mRNA and protein, and involves control of transcription, mRNA stability, post-translational modifications, cellular trafficking and compartmentalization, protein turnover and targeting to the *Ig* loci (Fig. 6.5) (reviewed in Refs. 31 to 33).

Regulation at gene and mRNA level: Induction
and transcriptional/post-transcriptional regulation

AID is expressed in a cell- and differentiation stage-specific manner (reviewed in Ref. 34). It is most highly expressed in B cells activated to undergo SHM and/or CSR, which *in vivo* are the antigen-activated

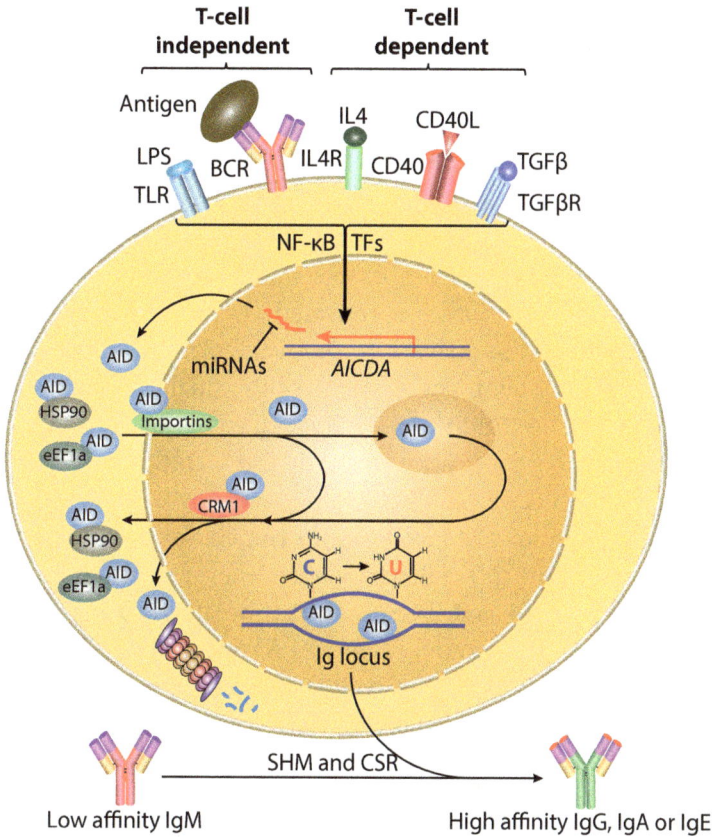

Fig. 6.5. Regulation of AID expression and subcellular localization in B cells. Transcription of the *AICDA* gene depends on NF-κB and other transcription factors (TF) and is activated after stimulation by antigens and cytokines. AICDA mRNA stability and translational efficiency are regulated by several miRNAs. AID protein is stabilized and retained in cytoplasm by interaction with chaperone HSP90 and/or translational elongation factor eEF1a. AID is transported to the nucleus by the importin pathway. In the nucleus AID transits nucleoli and accumulate in other subnuclear domains which are associated with splicing of mRNA. Nuclear AID is unstable due to both ubiquitin-dependent and ubiquitin-independent proteasomal degradation. In addition, AID is actively exported out of the nucleus by the CRM1 exportin. At the Ig loci AID deaminates cytosine to uracil and initiates SHM and CSR resulting in B cells expressing high affinity antibodies with new isotypes. See text for details.

germinal center B cells in secondary lymphoid organs. Transcription of the AID gene is regulated by four major conserved regulatory regions; region 1 (promoter), region 2 (enhancer and silencer in intron), region 3 (regulatory element 6 kb downstream from the last exon 5) and region 4 (enhancer 8 kb upstream from exon 1). Gene expression depends on activation of NF-κB, which binds directly to the gene promoter (region 1). Both the canonical (p65) and non-canonical (p52) subunits of NF-κB are important to initiate and sustain transcription, and both are activated by primary CSR-inducing stimuli.[35] One such stimulus is LPS that can turn on AID expression by a T cell-independent pathway, through dual activation of toll like receptor 4 (TLR4) and the B-cell receptor (BCR). By contrast, the T cell-dependent pathway goes through the CD40 ligand (CD40L) and direct T cell–B cell contact. CD40L, which is primarily expressed on the cell membrane of T_{FH} cells, binds the CD40 receptor on the B cell and activates both NF-κB pathways. Other T cell-derived factors such as cytokines IL-4 and TGF-β also stimulate AID expression by activating transcription factors, such as PAX5, STAT6, IRF4, CEBPs and E proteins that bind to the regulatory gene regions. In addition to its role in inducing AID expression, cytokines like IL-4 and TGF-β also initiate transcription of internal Ig switch regions, which are essential in CSR.

In addition to NF-κB, which is ubiquitously expressed and regulates many genes, B cell-specific and differentiation stage-specific transcription factors are involved in regulating the induction of AID. One such factor is HOXC4. HOXC4 is preferentially expressed in germinal center B cells and is upregulated by factors that stimulate AID expression and CSR (LPS, CD40L, IL4). It binds to a highly conserved site in the *AICDA* (and *Aicda*) promoter, and HOXC4 deficiency lower AID expression and impairs CSR and SHM.[36]

In non-B cells and naïve B cells, AID transcription is generally subject to negative regulation through several repressive regulatory elements that bind ubiquitous silencers such as MYB and E2F.[37] Furthermore, in activated B cells AID expression is only transient. When the BCR has achieved high affinity to antigen, the gene is turned off to stop further Ig diversification. The cells then differentiate into antibody-secreting plasma cells and memory cells. One mechanism involved in this is mediated thorough BCR signaling that downregulates AID gene expression through Ca^{2+}/calmodulin-dependent inhibition of E2A.[38,39] Other negative modulators of transcription are BLIMP1 and ID2, which are highly expressed in plasma and memory B cells and take part in silencing the *AICDA* gene.[40] Additional factors that are involved in modulating AID expression are the sex hormones and functional estrogen receptors that are expressed during B cell development. The *AICDA* gene has several binding sites for the estrogen receptor that activates transcription, while progesterone has opposite effect.[41,42] It is well established that sex hormones have an impact on adaptive immunity.[43] However, whether this is related to differential expression of AID is not known.

Infection by certain virus and bacteria as well as chronic inflammation can induce AID expression in naïve B cells as well as in non-B cells, such as hepatocytes and epithelial cells. This may in fact be a contributing factor to inflammation-induced oncogenesis.[44] Finally, AID expression is regulated post-transcriptionally by several microRNAs (miRs). miR155, miR181b and miR93 bind to conserved complementary sequences in the 3'-untranslated region of the *AICDA* mRNA, resulting in targeted mRNA degradation and/or translational repression.[45-47] Thus, balancing the intricate network of stimulating and inhibiting factors that regulate transcription of the *AICDA* gene and its mRNA stability is important to restrict AID expression to activated germinal center B cells.

Regulation of the AID protein: cellular localization, trafficking and protein turnover

After the protein has been expressed in the cytosol, AID needs to enter the nucleus to access the *Ig* genes. The initial observation that AID localizes predominantly to the cytoplasm of hypermutating B cells was therefore somewhat unexpected.[48] On the other hand, due to the mutagenic nature of the AID enzyme and its small size (small enough to diffuse trough the nuclear pore), it is reasonable that cells have strategies to limit its access to the nucleus and the genome. It has now been established that several mechanisms are involved in regulating the equilibrium between nuclear and cytoplasmic localization of AID. The enzyme shuttles in and out of the nucleus (Fig. 6.5). Outward transport is mediated by nuclear exportin CRM1 that binds to a classical nuclear export signal in the AID C-terminal region.[49,50] This C-terminal region also mediates retention of AID in the cytoplasm by associating with a large complex containing the translation elongation factor eEF1A1.[51-53] To overcome the cytoplasmic retention, AID is actively transported to the nucleus by nuclear importins (Fig. 6.5); which bind a positively charged surface motif on AID formed by both N-and C-terminal residues.[52,54] This motif also directs AID to nucleolus where it co-localizes with several of its interaction partners.[54] Subnuclear trafficking of AID is likely very important for its physiological function. In addition to its transit through nucleoli, AID also localizes to Cajal bodies and nuclear speckles, which are subnuclear domains enriched in RNA-processing and splicing factors.[55] Interestingly, this localization of AID is in accordance with recent findings demonstrating that splicing is involved in AID targeting to *Ig* switch regions.[56] Moreover, AID localization is also affected by the protein turnover since AID is more stable in the cytoplasm than in the nucleus.[57] In the cytoplasm AID is stabilized by chaperone HSP90,[58] while in the nucleus AID is rapidly degraded by both ubiquitin-dependent and ubiquitin-independent pathways.[57,59] Finally, the cell cycle seems to plays a role in regulating the access of AID to its substrate DNA. AID is transiently more localized to the nucleus in early G1 phase, coinciding with accumulation of uracil at the *Ig* genes.[14]

Altogether, cytoplasmic retention, active nuclear import and export, subnuclear trafficking and compartmentalization, proteosomal degradation and cell cycle-dependent mechanisms take part in the complex regulation of AIDs access to the nucleus and chromatin. These intricate control mechanisms of cellular localization probably contribute to protect the genome from untargeted mutations and permit the right number of AID molecules to access DNA. However, these mechanisms do not explain how AID is targeted to specific regions of the *Ig* genes.

6.1.3. *Targeting of AID to Ig Loci*

It was known prior to the discovery of AID that mutations in the *Ig* variable regions were closely associated with the gene transcription start sites. Cloning of the *Ig* promoter upstream of the *Igκ* constant region, which is normally not mutated, resulted in mutations at variable and constant regions with similar frequency and hallmarks.[60] This suggested that SHM was linked to transcription. This is also the case for CSR. The mouse and human *IgH* loci are similarly organized, with S-regions (1-12 kb) preceding each C-region (Fig. 6.3). Except from the Sμ region that is constitutively expressed, each S-region has its own upstream activation/cytokine-inducible promoter (intronic enhancer) that regulates transcription of a non-coding transcript. Activation of isolated splenic mouse B cells with LPS induces transcription of Sγ2b and Sγ3 regions and promotes CSR to IgG2b and IgG3. Likewise, cytokine IL4 activates Sγ1 and Sε transcription and switching to IgG1 and IgE, whereas TGF-β activates Sα transcription and allows switching to IgA. The correlation between activation of transcription of S-regions and switching to the corresponding isotype provides a strong mechanistic link between the two processes (reviewed in[61]). After the discovery of AID it was shown that AID-mediated mutagenesis in bacteria requires transcription, and that the mutation frequency was directly proportional to the transcription rate of the target gene.[62] Dependence of transcription for AID targeting has since been well documented in several systems (reviewed in[63]), and is crucial for AIDs access to its single stranded DNA substrate.[26,64] Recruitment of AID to single-stranded DNA is also facilitated by its interaction with the single-strand DNA-binding heterotrimer replication protein A (RPA).[65]

The association of AID with the transcription machinery is dependent on the elongation factor SPT5, which associates with stalled or paused RNA polymerase II.[66] Stalling of the transcription complex probably gives the relatively inefficient AID enzyme more time to deaminate its DNA substrate. AID also interacts with several components of the RNA exosome, which degrades transcripts that are hybridized to the DNA template and gives AID access to both strands in transcribed duplex DNA.[67]

A functional link between CSR and splicing was also suggested prior to the discovery of AID. It was shown that removing the splicing ability of the Sμ non-coding transcript prevented CSR.[68,69] Today we know that AID is physically linked to the splicing machinery. AID localizes to subnuclear domains, such as Cajal bodies and nuclear speckles, and physically associates with important splicing factors.[55] Thus, AID is likely targeted to the *Ig* loci by a transcription-coupled splicing-mediated mechanism. Recently, a direct functional link between splicing of the switch transcript, AID targeting and CSR was demonstrated. These results suggest that the S-region intron splice lariats are processed and fold into G-quadruplex structures that bind to the AID protein. Here it acts as specific RNA guide molecules that targets AID to *Ig* switch regions.[56] Interestingly, this is analogous to bacterial adaptive immune system CRISPR.[70]

Many additional factors are likely involved in recruiting AID to the *Ig* loci. AID seems to be part of multiprotein complexes containing the transcription machinery, transcription modulators, chromatin remodelers, RNA processing and splicing factors. A small variation in composition of these large complexes may be important for targeting to *IgV* or S-regions, which differ by topology and sequence. A comprehensive characterization of the complete AID interactome will contribute to unraveling the AID targeting mechanisms in the future.

6.1.4. *Processing of Uracil at the Ig Loci Mediates SHM and CSR*

Processing of the U:G mismatches generated by AID are essential to continue the antibody diversification processes. Gene knockout mouse models and human patients with inactivating mutations in the *UNG* gene have demonstrated that this is performed by the uracil-DNA glycosylase UNG.[9–12] UNG is required for normal CSR efficiency and also modulates the mutation spectrum at *Ig* variable regions during SHM. In addition to UNG, the MSH2/MSH6 mismatch sensor is involved in both processes. While CSR and SHM are perturbed by single deficiency in either UNG or MSH2, the combined double knockout mice are completely CSR deficient and display no SHM at A:T sites.[71,72] Normally, UNG and the MSH2/MSH6 mismatch recognition complex act in separate error-free post replicative DNA repair pathways during S phase of the cell cycle. By contrast, when acting in concert upon AID-induced uracils in activated B cells, they introduce mutations and DNA breaks at the Ig variable and switch regions. How this switch from faithful to error-prone uracil processing is regulated is not well understood. However, it may at least in part result from a local high density of lesions. In addition, diversification mechanisms and error-free canonical repair seem to occur in different phases of the cell cycle.

Uracil processing pathways facilitating SHM

SHM requires transcription through the *Ig* variable exons, and the mutational pathway is initiated by AID deamination in a region starting ~100 bp downstream from the promoter and proceeding 1–2 kb into the gene body of both *Ig* heavy and light chain genes. Downstream of AID, several uracil processing pathways interact and operate in concert to generate mutational diversity (Fig. 6.6), but exactly how this is coordinated is not known. What we do know is that a fraction of the AID-introduced uracils is excised by UNG2 to generate AP sites, while some U:G sites are recognized by the MSH2/MSH6 mismatch sensor. UNG-deficiency results in faster accumulation of *Ig* mutations,[73] suggesting that some uracils at the *Ig* loci are faithfully repaired by UNG-initiated BER. Finally, some uracil sites may escape processing before replication, ending up as direct AID deamination footprint signatures (C to T transitions at AID hotspot context) (Fig. 6.6). The fraction of such transitions relative to transversions at *IgV* cytosine sites is strongly increased by UNG deficiency in both human and mouse B cells.[10,11] This indicates that AP sites generated by UNG at the site of deamination are bypassed by specialized polymerases that introduce transversion mutations. The polymerase REV1 that incorporates one dCMP opposite an abasic site, is responsible for most of these mutations.[74] In addition, a mutagenic

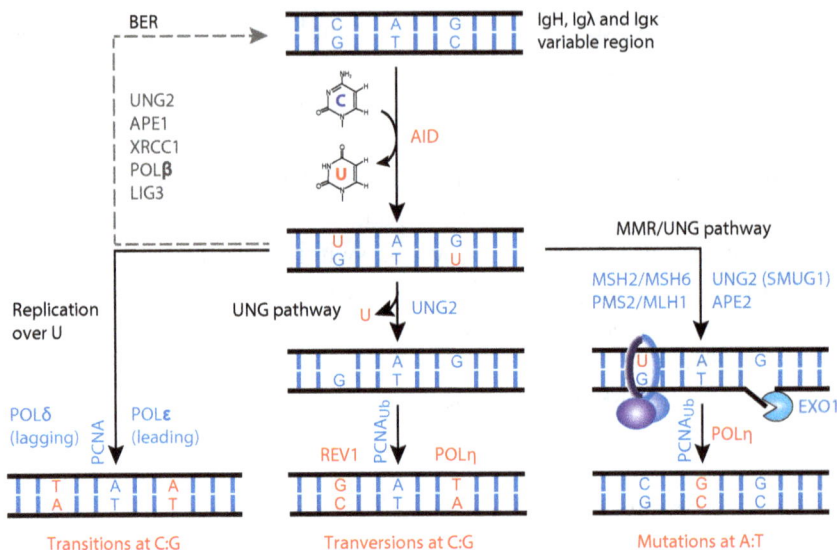

Fig. 6.6. Somatic hypermutation pathways. Generation of mutations during SHM can be explained by three main molecular pathways. 1. Direct replication pathway: AID generated uracil results in C to T transitions. 2. UNG excision pathway: UNG2 excises the uracil leaving an AP site that is bypassed by REV1. The AP site may alternatively be processed by error prone BER involving POLη. 3. MMR/UNG pathway: MSH2/MSH6 recognizes the U:G mismatch, PMS2/MLH1 and UNG2/APE2 generates EXO1 entry sites. POLη fills in the gaps.

version of BER involving specialized polymerases (POLη, REV1 and others?) instead of POLβ may also contribute downstream from uracil excision in the UNG SHM-pathway.

Importantly, approximately half of the mutations generated during SHM are not at C:G sites within AID hotspot motifs. These are A:T mutations that occur at a distance from the original deamination sites and depend on the mismatch repair factors MSH2/MSH6 (MutSα) and exonuclease 1 (EXO1), together with the DNA polymerase POLη (Fig. 6.6). Knocking out any of these factors decreases mutations at A:T sites (reviewed in[75]). Another important event is monoubiquitination of PCNA at lysine 164 (PCNA$_{Ub}$), which allows recruitment of the translesion polymerase POLη.[76,77] In accordance with this, mice expressing PCNA mutated at lysine 164 show low levels of *Ig* mutations at A:T pairs.[78]

To make this even more complicated, the mutagenic version of MMR that operates during SHM, cooperates with UNG2 and AP endonuclease (likely APE2) for generation of nicks that can serve as entry sites for EXO1. Although UNG deficiency alone does not significantly affect A:T mutations,[10] double knockouts of UNG and MSH2 completely ablate all A:T mutations.[71] Normally, the uracil-DNA glycosylase SMUG1 does not seem to have a role in Ig diversification. However, SMUG1 may act as a backup mechanism in the absence of UNG to generate the nick entry sites for the exonuclease, since *Ung*[-/-]*Smug1*[-/-] double knockout mice display a significant reduction of mutations at A:T sites in contrast to the single knockouts.[79] In germinal center B cells, APE1 expression, the major AP-endonuclease in BER, is strongly downregulated. Concomitantly, the much less efficient homolog APE2 is highly upregulated, and APE2 deficient mice show a specific reduction in A:T mutations,[80,81] indicating that APE2 is the active AP endonuclease in SHM.

The MMR-nicking complex PMS2/MLH1 (MutLα) has been considered dispensable for SHM and A:T mutagenesis. This is based on the observation that these processes are not significantly affected in PMS2-deficient mice.[75] Recently, however, *Pms2*[-/-] mice were crossed with Ung[-/-] mice, and the double knockouts displayed a 50% reduction in SHM at A:T base pairs, with the remaining A:T mutations depending on the uracil-DNA glycosylases SMUG1 and TDG.[82] This suggests that the PMS2/MLH1 nicking complex and the uracil-DNA glycosylases/APE pathway work in parallel to generate nicks for EXO1. Moreover, the study indicates that the uracil-DNA glycosylase/APE pathway acts mainly on the coding strand, while PMS2/MLH1 does not display any strand bias.[82]

Taken together, this demonstrates that BER factors cooperate with MMR factors and POLη to generate mutations at A:T sites. Thus, during SHM, uracil processing factors from different pathways are orchestrated in new ways at the *Ig* variable loci to facilitate mutations instead of repair.

Switch region topology and uracil processing pathways that facilitate double strand breaks and CSR

As described above, the *IgH* locus has S-regions upstream of each isotype-specific C-region (Fig. 6.3). The topology of the S-regions is important for CSR as they serve as the recombination targets. The S-regions are 1–12 kb long and consist of repetitive motifs, and there is a direct correlation between S-region length and CSR frequency.[83] Mouse Sμ, for example, is 3.2 kb long packed with GAGCT pentamer motifs, containing palindromic AID hotspot sequences. Thus, S-regions have likely evolved to become the perfect DNA substrate for AID and generation of clustered deaminations that by uracil processing ends up as ds breaks. Although the results are different, the molecular mechanisms that generate ds breaks during CSR are related to the SHM pathways (Fig. 6.6). Uracil processing in switch regions can be divided in two main pathways; the UNG pathway and the connected MMR/UNG pathway (Fig. 6.7).

The UNG pathway depends on clustered deamination events in both strands, for example in the palindromic AGCT/AGCT hotspot sequence. Uracil excisions followed by strand incisions by AP endonuclease will generate ds breaks directly. Ape1[+/−] heterozygous mice display decreased CSR,[84] while APE2-deficient mice have normal switching efficiency.[81] Thus, the major AP endonuclease APE1 is likely involved in this step. Interestingly, this is in contrast to the SHM A:T mutagenesis pathway where APE2 play a role.[80,81]

When the deamination events take place at more distant sites (may be several kb apart), UNG2/APE1 and components of the MMR pathway cooperate to generate ds breaks. This scenario can be as follows: AID-induced uracil is processed conventionally by UNG2 and APE1 in the upper strand, resulting in a single-strand nick (Fig. 6.7). A second distant U:G mismatch, having uracil on the opposite lower strand, is recognized by MSH2/MSH6 that recruits the MMR nicking complex MLH1/PMS2. The endonuclease activity of PMS2 then generates nicks in DNA that can serve as entry cites for the exonuclease EXO1. Collision of the EXO1 digestion tract in the lower strand with the UNG2/APE1-generated nick in the upper strand will result in ds breaks. This suggested pathway is supported by

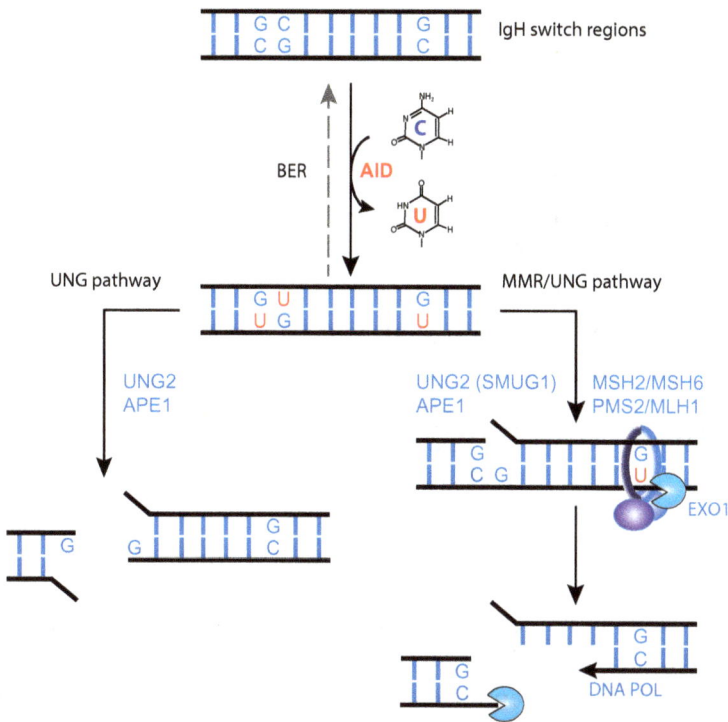

Fig. 6.7. Generation of double strand breaks in Ig switch regions that initiate CSR. Uracil processing in switch regions can generate ds breaks by two main pathways. The UNG pathway relies on closely localized uracils on opposite strands. APE1 incises at the AP sites generated by UNG. In the MMR/UNG pathway distantly spaced uracil can be processed to ds breaks. MSH2/MSH6 recognizes the U:G mismatch. MLH1/PMS2 is then recruited and generates nicks that can serve as entry sites for EXO1. EXO1 degrades the lower strand until it meets an UNG2/ APE1 generated nick at the upper strand. SMUG1 may serve as a backup uracil excision activity when UNG is absent.

mouse models. Mice that are deficient for MSH2, MSH6, MLH1, PMS2 or EXO1, all exhibit a 50% or greater reduction in switching efficiency in splenic B cells stimulated *ex vivo,* or they display low levels of switched Ig isotypes in serum *in vivo.*[75]

UNG is central in both CSR pathways outlined (Fig. 6.7), but the CSR deficiency in *Ung*[-/-] mice is not complete as they build up switched Ig isotypes in serum over time.[10] In the same way as SMUG1 seem to have a backup role for UNG during SHM, most of the residual class switching in *Ung*[/-] mice depends on the SMUG1 uracil-DNA glycosylase.[79] In accordance with the observed species difference between man and mouse in genomic uracil processing,[85] a backup role of SMUG1 in adaptive immunity seem to be more pronounced in mice than in human UNG-deficient patients, which display a more severe CSR deficient phenotype.[11] Taken together this demonstrates that uracil excision is an essential requirement to generate the ds breaks in switch regions and promote class switching.

6.2. Innate Immunity; Uracil in DNA is an Intermediate in Restriction of Viral Pathogens

While AID is the key enzyme in adaptive immunity, other enzymes in the APOBEC polynucleotide deaminase family perform important functions in the innate immune system. The APOBEC3 (A3) enzymes are present in all placental mammals where they play an important role in restricting replication of retroviruses and retrotransposons. Like AID, they are single-stranded DNA cytosine deaminases that uracilates its substrates. The A3 gene array on human chromosome 22 includes seven functional genes.[86] It likely arose by gene duplication of a single-copy primordial gene and encodes the A3 family members A3A, A3B, A3C, A3DE, A3F, A3G and A3H. Interestingly, depending on the species, the A3 genes have expanded and/or contracted. The A3 gene number ranges from one in mice, rats and pigs, three in cats, six in horse, and seven in humans and most primates. In addition, the human A3 genes are highly polymorphic, likely due to selective pressure and positive selection during primate evolution.[7]

The same year that the A3 genes were identified, A3G (also previously called CEM15) was shown to be an HIV-1 restriction factor that inhibits viral replication.[87] Since then it has been demonstrated that many of the A3 cytidine deaminases affect the replication of a variety of viruses and intracellular retrotransposons (reviewed in Refs. 1, 88 and 89). The A3 enzymes are most highly expressed in immune cells but they are present also in other human cells types and tissues.[90,91] They are induced by mediators of inflammation, which likely reflects their role as a first line defense against invading viruses. Intrinsic specificities of A3 proteins together with expression level in various cell types likely determine the impact of each family member on viral restriction.

The physiological function of the A3 cytidine deaminases has been best studied in relation to HIV-1 restriction in T cells, and a model is shown in Fig. 6.8. Here A3 proteins (A3DE, A3G, A3F and A3H), which are expressed in the cytoplasm of virus infected cells, are

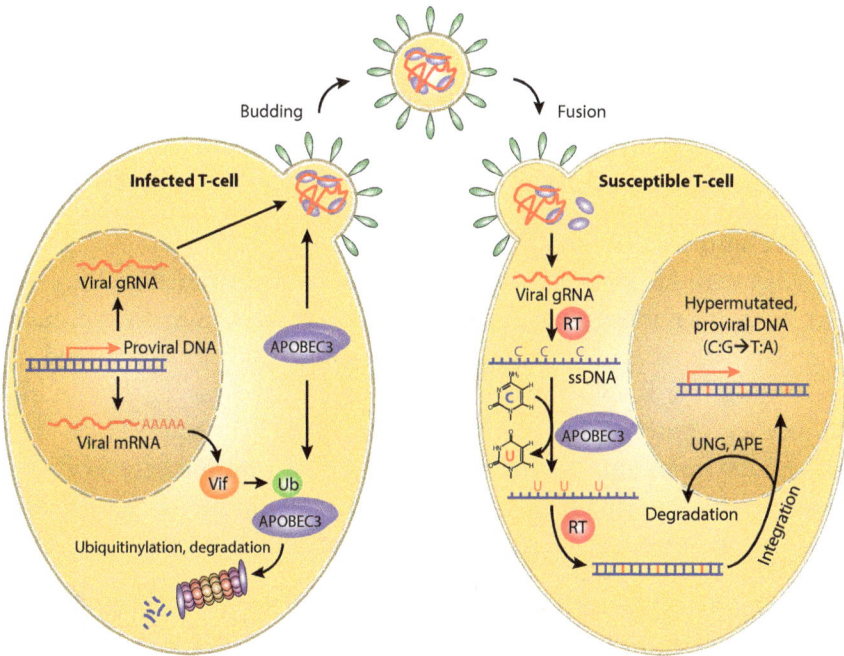

Fig. 6.8. Model of virus restriction by APOBEC3 (A3) enzymes and contradiction of this defence system by the HIV-1 Vif protein. HIV-1 infection of T cells induces expression of A3 enzymes. The viruses overcome this threat by expressing the Vif protein that induces ubiquitin-dependent degradation of A3. In absence of Vif the A3 proteins are incorporated together with the viral genomic RNA into budding virions. During subsequent infection of susceptible T cells A3 enzymes are released and deaminate viral cDNA after reverse transcription of the viral RNA. This results in hypermutation and/or degradation (by UNG/APE) of viral DNA.

packed into the HIV-1 virions, likely by binding to viral genomic RNA.[92] In susceptible target cells, viral RNA and APOBECs are released and a single stranded cDNA intermediate is generated from the viral genome by reverse transcriptase (RT). On this ss DNA intermediate, the A3s deaminate cytosine to uracil. This results in C:G to T:A hypermutation and/or degradation of the proviral DNA by UNG (or SMUG1) and APE (Fig. 6.8).

Successful viruses have evolved sophisticated immune suppression strategies to counteract this host defense mechanism. To neutralize viral restriction by the A3 deaminases, HIV-1 and other viruses express the Vif (Virion infectivity factor) protein. Vif is a small basic protein that binds to A3 enzymes and recruits a multiprotein E3 ubiquitin ligase complex that ubiquitylates and targets A3s for proteasomal degradation.[93,94] In addition, another HIV protein, Vpr (Viral protein R) binds UNG and SMUG1 and induces their degradation, also this by an ubiquitin-dependent pathway.[95] Thus, manipulating A3 expression to override the antagonistic activity of Vif or inhibition of Vif and/or Vpr might constitute a future approach to fight viral infection.

AID and A3 DNA deaminases have important functions in protecting us from infection and disease. They act both in first line innate immune defense by restricting viral replication and spreading, and in the second line adaptive immune system that develops highly specific antibodies against different types of invading pathogens. Thus, our immune system utilizes uracil in DNA as an intermediate both to directly mutate and degrade viruses and to develop specific antibodies including immunological memory. Although DNA cytosine deamination has obvious benefits for the organism, these processes may also cause untargeted mutations that may lead to cancer. We know that untargeted mutations by AID are involved in development of B-cell lymphoma.[33] Moreover, while AID is expressed almost exclusively in activated B cells, the A3 enzymes have much broader expression profiles that span most tissues in the human body.[91] In accordance with this, a mutational signature present in more than half of examined cancer types has been attributed to the A3 family of DNA deaminases.[96,97] Taken together, this demonstrates that the DNA-uracilation strategy used by our immune system has many benefits but also dangers. These dangers will be more extensively covered in the next chapter.

References

1. Harris, R. S., Dudley, J. P. APOBECs and virus restriction. *Virology* **479–480**, 131–145 (2015).
2. Carmona, L. M., Schatz, D. G. New insights into the evolutionary origins of the recombination-activating gene proteins and V(D)J recombination. *FEBS J* (2016).
3. Muramatsu, M., Sankaranand, V. S., Anant, S., *et al.* Specific expression of activation-induced cytidine deaminase (AID), a novel member of the RNA-editing deaminase family in germinal center B cells. *J Biol Chem* **274**, 18470–18476 (1999).
4. Muramatsu, M., Kinoshita, K., Fagarasan, S., *et al.* Class switch recombination and hypermutation require activation-induced cytidine deaminase (AID), a potential RNA editing enzyme. *Cell* **102**, 553–563 (2000).
5. Revy, P., Muto, T., Levy, Y., *et al.* Activation-induced cytidine deaminase (AID) deficiency causes the autosomal recessive form of the Hyper-IgM syndrome (HIGM2). *Cell* **102**, 565–575 (2000).
6. Durandy, A., Kracker, S. Immunoglobulin class-switch recombination deficiencies. *Arthritis Res Ther* **14**, 218 (2012).
7. Conticello, S. G., Thomas, C. J., Petersen-Mahrt, S. K., Neuberger, M. S. Evolution of the AID/APOBEC family of polynucleotide (deoxy)cytidine deaminases. *Mol Biol Evol* **22**, 367–377 (2005).
8. Teng, B., Burant, C. F., Davidson, N. O. Molecular cloning of an apolipoprotein B messenger RNA editing protein. *Science* **260**, 1816–1819 (1993).
9. Di Noia, J., Neuberger, M. S. Altering the pathway of immunoglobulin hypermutation by inhibiting uracil-DNA glycosylase. *Nature* **419**, 43–48 (2002).
10. Rada, C., Williams, G. T., Nilsen, H., *et al.* Immunoglobulin isotype switching is inhibited and somatic hypermutation perturbed in UNG-deficient mice. *Curr Biol* **12**, 1748–1755 (2002).
11. Imai, K., Slupphaug, G., Lee, W. I., *et al.* Human uracil-DNA glycosylase deficiency associated with profoundly impaired immunoglobulin class-switch recombination. *Nat Immunol* **4**, 1023–1028 (2003).
12. Kavli, B., Andersen, S., Otterlei, M., *et al.* B cells from hyper-IgM patients carrying UNG mutations lack ability to remove uracil from ssDNA and have elevated genomic uracil. *J Exp Med* **201**, 2011–2021 (2005).

13. Maul, R. W., Saribasak, H., Martomo, S. A., *et al.* Uracil residues dependent on the deaminase AID in immunoglobulin gene variable and switch regions. *Nat Immunol* **12**, 70–76 (2011).

14. Wang, Q., Kieffer-Kwon, K. R., Oliveira, T. Y., *et al.* The cell cycle restricts activation-induced cytidine deaminase activity to early G1. *J Exp Med* **214**, 49–58 (2017).

15. Di Noia, J. M., Neuberger, M. S. Molecular mechanisms of antibody somatic hypermutation. *Annu Rev Biochem* **76**, 1–22 (2007).

16. Kohli, R. M., Maul, R. W., Guminski, A. F., *et al.* Local sequence targeting in the AID/APOBEC family differentially impacts retroviral restriction and antibody diversification. *J Biol Chem* **285**, 40956–40964 (2010).

17. Pham, P., Afif, S. A., Shimoda, M., *et al.* Structural analysis of the activation-induced deoxycytidine deaminase required in immunoglobulin diversification. *DNA Repair (Amst)* **43**, 48–56 (2016).

18. Shi, K., Carpenter, M. A., Banerjee, S., *et al.* Structural basis for targeted DNA cytosine deamination and mutagenesis by APOBEC3A and APOBEC3B. *Nat Struct Mol Biol* **24**, 131–139 (2017).

19. Liu, M., Duke, J. L., Richter, D. J., *et al.* Two levels of protection for the B cell genome during somatic hypermutation. *Nature* **451**, 841–845 (2008).

20. Robbiani, D. F., Nussenzweig, M. C. Chromosome translocation, B cell lymphoma, and activation-induced cytidine deaminase. *Annu Rev Pathol* **8**, 79–103 (2013).

21. Pettersen, H. S., Galashevskaya, A., Doseth, B., *et al.* AID expression in B-cell lymphomas causes accumulation of genomic uracil and a distinct AID mutational signature. *DNA Repair (Amst)* **25**, 60–71 (2015).

22. Shalhout, S., Haddad, D., Sosin, A., *et al.* Genomic uracil homeostasis during normal B cell maturation and loss of this balance during B cell cancer development. *Mol Cell Biol* **34**, 4019–4032 (2014).

23. Bhutani, N., Brady, J. J., Damian, M., *et al.* Reprogramming towards pluripotency requires AID-dependent DNA demethylation. *Nature* **463**, 1042–1047 (2010).

24. Cortellino, S., Xu, J., Sannai, M., *et al.* Thymine DNA glycosylase is essential for active DNA demethylation by linked deamination-base excision repair. *Cell* **146**, 67–79 (2011).

25. Kumar, R., DiMenna, L., Schrode, N., *et al.* AID stabilizes stem-cell phenotype by removing epigenetic memory of pluripotency genes. *Nature* **500**, 89–92 (2013).

26. Bransteitter, R., Pham, P., Scharff, M. D., Goodman, M. F. Activation-induced cytidine deaminase deaminates deoxycytidine on single-stranded DNA but

requires the action of RNase. *Proc Natl Acad Sci U S A* **100**, 4102–4107 (2003).

27. Larijani, M., Frieder, D., Sonbuchner, T. M., *et al.* Methylation protects cytidines from AID-mediated deamination. *Mol Immunol* **42**, 599–604 (2005).

28. Schuermann, D., Weber, A. R., Schar, P. Active DNA demethylation by DNA repair: Facts and uncertainties. *DNA Repair (Amst)* **44**, 92–102 (2016).

29. Rogozin, I. B., Lada, A. G., Goncearenco, A., *et al.* Activation induced deaminase mutational signature overlaps with CpG methylation sites in follicular lymphoma and other cancers. *Sci Rep* **6**, 38133 (2016).

30. Ramiro, A. R., Barreto, V. M. Activation-induced cytidine deaminase and active cytidine demethylation. *Trends Biochem Sci* **40**, 172–181 (2015).

31. Xu, Z., Zan, H., Pone, E. J., *et al.* Immunoglobulin class-switch DNA recombination: induction, targeting and beyond. *Nat Rev Immunol* **12**, 517–531 (2012).

32. Orthwein, A., Di Noia, J. M. Activation induced deaminase: how much and where? *Semin Immunol* **24**, 246–254 (2012).

33. Casellas, R., Basu, U., Yewdell, W. T., *et al.* Mutations, kataegis and translocations in B cells: understanding AID promiscuous activity. *Nat Rev Immunol* **16**, 164–176 (2016).

34. Zan, H., Casali, P. Regulation of Aicda expression and AID activity. *Autoimmunity* **46**, 83–101 (2013).

35. Pone, E. J., Zhang, J., Mai, T., *et al.* BCR-signalling synergizes with TLR-signalling for induction of AID and immunoglobulin class-switching through the non-canonical NF-kappaB pathway. *Nat Commun* **3**, 767 (2012).

36. Park, S. R., Zan, H., Pal, Z., *et al.* HoxC4 binds to the promoter of the cytidine deaminase AID gene to induce AID expression, class-switch DNA recombination and somatic hypermutation. *Nat Immunol* **10**, 540–550 (2009).

37. Tran, T. H., Nakata, M., Suzuki, K., *et al.* B cell-specific and stimulation-responsive enhancers derepress Aicda by overcoming the effects of silencers. *Nat Immunol* **11**, 148–154 (2010).

38. Hauser, J., Sveshnikova, N., Wallenius, A., *et al.* B-cell receptor activation inhibits AID expression through calmodulin inhibition of E-proteins. *Proc Natl Acad Sci U S A* **105**, 1267–1272 (2008).

39. Hauser, J., Verma-Gaur, J., Wallenius, A., Grundstrom, T. Initiation of antigen receptor-dependent differentiation into plasma cells by calmodulin inhibition of E2A. *J Immunol* **183**, 1179–1187 (2009).

40. Xu, Z., Pone, E. J., Al-Qahtani, A., *et al.* Regulation of aicda expression and AID activity: relevance to somatic hypermutation and class switch DNA recombination. *Crit Rev Immunol* **27**, 367–397 (2007).

41. Pauklin, S., Sernandez, I. V., Bachmann, G., *et al.* Estrogen directly activates AID transcription and function. *J Exp Med* **206**, 99–111 (2009).
42. Pauklin, S., Petersen-Mahrt, S. K. Progesterone inhibits activation-induced deaminase by binding to the promoter. *J Immunol* **183**, 1238–1244 (2009).
43. Grimaldi, C. M., Hill, L., Xu, X., *et al.* Hormonal modulation of B cell development and repertoire selection. *Mol Immunol* **42**, 811–820 (2005).
44. Marusawa, H., Takai, A., Chiba, T. Role of activation-induced cytidine deaminase in inflammation-associated cancer development. *Adv Immunol* **111**, 109–141 (2011).
45. Teng, G., Hakimpour, P., Landgraf, P., *et al.* MicroRNA-155 is a negative regulator of activation-induced cytidine deaminase. *Immunity* **28**, 621–629 (2008).
46. de Yebenes, V. G., Belver, L., Pisano, D. G., *et al.* miR-181b negatively regulates activation-induced cytidine deaminase in B cells. *J Exp Med* **205**, 2199–2206 (2008).
47. Basso, K., Schneider, C., Shen, Q., Holmes *et al.* BCL6 positively regulates AID and germinal center gene expression via repression of miR-155. *J Exp Med* **209**, 2455–2465 (2012).
48. Rada, C., Jarvis, J. M., Milstein, C. AID-GFP chimeric protein increases hypermutation of Ig genes with no evidence of nuclear localization. *Proc Natl Acad Sci U S A* **99**, 7003–7008 (2002).
49. Ito, S., Nagaoka, H., Shinkura, R., *et al.* Activation-induced cytidine deaminase shuttles between nucleus and cytoplasm like apolipoprotein B mRNA editing catalytic polypeptide 1. *Proc Natl Acad Sci U S A* **101**, 1975–1980 (2004).
50. McBride, K. M., Barreto, V., Ramiro, A. R., *et al.* Somatic hypermutation is limited by CRM1-dependent nuclear export of activation-induced deaminase. *J Exp Med* **199**, 1235–1244 (2004).
51. Hasler, J., Rada, C., Neuberger, M. S. Cytoplasmic activation-induced cytidine deaminase (AID) exists in stoichiometric complex with translation elongation factor 1alpha (eEF1A). *Proc Natl Acad Sci U S A* **108**, 18366–18371 (2011).
52. Patenaude, A. M., Orthwein, A., Hu, Y., *et al.* Active nuclear import and cytoplasmic retention of activation-induced deaminase. *Nat Struct Mol Biol* **16**, 517–527 (2009).
53. Methot, S. P., Litzler, L. C., Trajtenberg, F., *et al.* Consecutive interactions with HSP90 and eEF1A underlie a functional maturation and storage pathway of AID in the cytoplasm. *J Exp Med* **212**, 581–596 (2015).
54. Hu, Y., Ericsson, I., Torseth, K., *et al.* A combined nuclear and nucleolar localization motif in activation-induced cytidine deaminase (AID) controls immunoglobulin class switching. *J Mol Biol* **425**, 424–443 (2013).

55. Hu, Y., Ericsson, I., Doseth, B., *et al.* Activation-induced cytidine deaminase (AID) is localized to subnuclear domains enriched in splicing factors. *Exp Cell Res* **322**, 178–192 (2014).

56. Zheng, S., Vuong, B. Q., Vaidyanathan, B., *et al.* Non-coding RNA Generated following Lariat Debranching Mediates Targeting of AID to DNA. *Cell* **161**, 762–773 (2015).

57. Aoufouchi, S., Faili, A., Zober, C., *et al.* Proteasomal degradation restricts the nuclear lifespan of AID. *J Exp Med* **205**, 1357–1368 (2008).

58. Orthwein, A., Patenaude, A. M., Affar el, B., *et al.* Regulation of activation-induced deaminase stability and antibody gene diversification by Hsp90. *J Exp Med* **207**, 2751–2765 (2010).

59. Uchimura, Y., Barton, L. F., Rada, C., Neuberger, M. S. REG-gamma associates with and modulates the abundance of nuclear activation-induced deaminase. *J Exp Med* **208**, 2385–2391 (2011).

60. Peters, A., Storb, U. Somatic hypermutation of immunoglobulin genes is linked to transcription initiation. *Immunity* **4**, 57–65 (1996).

61. Matthews, A. J., Zheng, S., DiMenna, L. J., Chaudhuri, J. Regulation of immuno-globulin class-switch recombination: choreography of noncoding transcription, targeted DNA deamination, and long-range DNA repair. *Adv Immunol* **122**, 1–57 (2014).

62. Ramiro, A. R., Stavropoulos, P., Jankovic, M., Nussenzweig, M. C. Transcription enhances AID-mediated cytidine deamination by exposing single-stranded DNA on the nontemplate strand. *Nat Immunol* **4**, 452–456 (2003).

63. Storb, U. Why does somatic hypermutation by AID require transcription of its target genes? *Adv Immunol* **122**, 253–277 (2014).

64. Dickerson, S. K., Market, E., Besmer, E., Papavasiliou, F. N. AID mediates hypermutation by deaminating single stranded DNA. *J Exp Med* **197**, 1291–1296 (2003).

65. Chaudhuri, J., Khuong, C., Alt, F. W. Replication protein A interacts with AID to promote deamination of somatic hypermutation targets. *Nature* **430**, 992–998 (2004).

66. Pavri, R., Gazumyan, A., Jankovic, M., *et al.* Activation-induced cytidine deaminase targets DNA at sites of RNA polymerase II stalling by interaction with Spt5. *Cell* **143**, 122–133 (2010).

67. Basu, U., Meng, F. L., Keim, C., *et al.* The RNA exosome targets the AID cytidine deaminase to both strands of transcribed duplex DNA substrates. *Cell* **144**, 353–363 (2011).

68. Lorenz, M., Jung, S., Radbruch, A. Switch transcripts in immunoglobulin class switching. *Science* **267**, 1825–1828 (1995).

69. Hein, K., Lorenz, M. G., Siebenkotten, G., *et al.* Processing of switch transcripts is required for targeting of antibody class switch recombination. *J Exp Med* **188**, 2369–2374 (1998).

70. Barrangou, R., Fremaux, C., Deveau, H., *et al.* CRISPR provides acquired resistance against viruses in prokaryotes. *Science* **315**, 1709–1712 (2007).

71. Rada, C., Di Noia, J. M., Neuberger, M. S. Mismatch recognition and uracil excision provide complementary paths to both Ig switching and the A/T-focused phase of somatic mutation. *Mol Cell* **16**, 163–171 (2004).

72. Xue, K., Rada, C., Neuberger, M. S. The in vivo pattern of AID targeting to immunoglobulin switch regions deduced from mutation spectra in msh2–/–ung–/– mice. *J Exp Med* **203**, 2085–2094 (2006).

73. Saribasak, H., Saribasak, N. N., Ipek, F. M., *et al.* Uracil DNA glycosylase disruption blocks Ig gene conversion and induces transition mutations. *J Immunol* **176**, 365–371 (2006).

74. Jansen, J. G., Langerak, P., Tsaalbi-Shtylik, A., *et al.* Strand-biased defect in C/G transversions in hypermutating immunoglobulin genes in Rev1-deficient mice. *J Exp Med* **203**, 319–323 (2006).

75. Zanotti, K. J., Gearhart, P. J. Antibody diversification caused by disrupted mismatch repair and promiscuous DNA polymerases. *DNA Repair (Amst)* **38**, 110–116 (2016).

76. Kannouche, P. L., Wing, J., Lehmann, A. R. Interaction of human DNA polymerase eta with monoubiquitinated PCNA: a possible mechanism for the polymerase switch in response to DNA damage. *Mol Cell* **14**, 491–500 (2004).

77. Guo, C., Sonoda, E., Tang, T. S., *et al.* REV1 protein interacts with PCNA: significance of the REV1 BRCT domain in vitro and in vivo. *Mol Cell* **23**, 265–271 (2006).

78. Langerak, P., Nygren, A. O., Krijger, P. H., *et al.* A/T mutagenesis in hypermutated immunoglobulin genes strongly depends on PCNAK164 modification. *J Exp Med* **204**, 1989–1998 (2007).

79. Dingler, F. A., Kemmerich, K., Neuberger, M. S., Rada, C. Uracil excision by endogenous SMUG1 glycosylase promotes efficient Ig class switching and impacts on A:T substitutions during somatic mutation. *Eur J Immunol* **44**, 1925–1935 (2014).

80. Stavnezer, J., Linehan, E. K., Thompson, M. R., *et al.* Differential expression of APE1 and APE2 in germinal centers promotes error-prone repair and A:T mutations during somatic hypermutation. *Proc Natl Acad Sci U S A* **111**, 9217–9222 (2014).

81. Sabouri, Z., Okazaki, I. M., Shinkura, R., *et al.* Apex2 is required for efficient somatic hypermutation but not for class switch recombination of immunoglobulin genes. *Int Immunol* **21**, 947–955 (2009).

82. Girelli Zubani, G., Zivojnovic, M., De Smet, A., *et al.* Pms2 and uracil-DNA glycosylases act jointly in the mismatch repair pathway to generate Ig gene mutations at A-T base pairs. *J Exp Med* (2017).

83. Zarrin, A. A., Tian, M., Wang, J., *et al.* Influence of switch region length on immunoglobulin class switch recombination. *Proc Natl Acad Sci U S A* **102**, 2466–2470 (2005).

84. Guikema, J. E., Linehan, E. K., Tsuchimoto, D., *et al.* APE1- and APE2-dependent DNA breaks in immunoglobulin class switch recombination. *J Exp Med* **204**, 3017–3026 (2007).

85. Doseth, B., Visnes, T., Wallenius, A., *et al.* Uracil-DNA glycosylase in base excision repair and adaptive immunity: species differences between man and mouse. *J Biol Chem* **286**, 16669–16680 (2011).

86. Jarmuz, A., Chester, A., Bayliss, J., *et al.* An anthropoid-specific locus of orphan C to U RNA-editing enzymes on chromosome 22. *Genomics* **79**, 285–296 (2002).

87. Sheehy, A. M., Gaddis, N. C., Choi, J. D., Malim, M. H. Isolation of a human gene that inhibits HIV-1 infection and is suppressed by the viral Vif protein. *Nature* **418**, 646–650 (2002).

88. Moris, A., Murray, S., Cardinaud, S. AID and APOBECs span the gap between innate and adaptive immunity. *Front Microbiol* **5**, 534 (2014).

89. Refsland, E. W., Harris, R. S. The APOBEC3 family of retroelement restriction factors. *Curr Top Microbiol Immunol* **371**, 1–27 (2013).

90. Koning, F. A., Newman, E. N., Kim, E. Y., Kunstman *et al.* Defining APOBEC3 expression patterns in human tissues and hematopoietic cell subsets. *J Virol* **83**, 9474–9485 (2009).

91. Refsland, E. W., Stenglein, M. D., Shindo, K., *et al.* Quantitative profiling of the full APOBEC3 mRNA repertoire in lymphocytes and tissues: implications for HIV-1 restriction. *Nucleic Acids Res* **38**, 4274–4284 (2010).

92. Bogerd, H. P., Cullen, B. R. Single-stranded RNA facilitates nucleocapsid: APOBEC3G complex formation. *RNA* **14**, 1228–1236 (2008).

93. Conticello, S. G., Harris, R. S., Neuberger, M. S. The Vif protein of HIV triggers degradation of the human antiretroviral DNA deaminase APOBEC3G. *Curr Biol* **13**, 2009–2013 (2003).

94. Desimmie, B. A., Delviks-Frankenberrry, K. A., Burdick, R. C., *et al.* Multiple APOBEC3 restriction factors for HIV-1 and one Vif to rule them all. *J Mol Biol* **426**, 1220–1245 (2014).

95. Schrofelbauer, B., Yu, Q., Zeitlin, S. G., Landau, N. R. Human immunodeficiency virus type 1 Vpr induces the degradation of the UNG and SMUG uracil-DNA glycosylases. *J Virol* **79**, 10978–10987 (2005).

96. Alexandrov, L. B., Nik-Zainal, S., Wedge, D. C., *et al.* Signatures of mutational processes in human cancer. *Nature* **500**, 415–421 (2013).

97. Petljak, M., Alexandrov, L. B. Understanding mutagenesis through delineation of mutational signatures in human cancer. *Carcinogenesis* **37**, 531–540 (2016).

Chapter 7

Genomic Uracil and Cancer

Antonio Sarno*, Pål Sætrom, Bodil Kavli and
Henrik Sahlin Pettersen

New technologies have made possible sequencing of a large number
of whole cancer genomes and exomes. Such studies have identified
sequence-associated mutational signatures in human cancers. These
frequently suggest a causative mutational agent, thus shedding light
on the molecular mechanisms of oncogenesis and cancer progression.
Analysis of mutational signatures may also be used to monitor
response to therapy.

Comprehensive studies of cancer genomes have indicated involvement of DNA-cytosine deaminases in oncogenesis. As described in
Chapter 6, such enzymes convert cytosines in DNA to uracils and
have preference for certain sequence contexts.[1–3] One study in particular analysed data from over 7,000 human cancer genomes and
found mutational signatures of DNA-cytosine deaminases in 16 of the
30 human cancer types studied, each usually having several genetic
subtypes.[2] Indeed, DNA-cytosine deaminase mutational signature
was the second most common, only after the aging signature (C to T
transitions at CpGs caused by deamination of 5-mC). Thus, U:G
mismatches constitute a major mutational DNA lesion, likely contributing to both cancer development and progression (reviewed in[4]).

* antonio.sarno@ntnu.no

Dysregulation of genomic uracil levels may lead to disease. That is, altering the "normal" amount of genomic uracil in a cell by increasing or decreasing its generation or repair may have deleterious consequences. In Chapter 6, we discussed the role of genomic uracil as an intermediate in adaptive immunity and the immunological pathogenicity of defective uracil processing in immunoglobulin genes during antibody affinity maturation. Either too much genomic uracil or too little may be detrimental in the adaptive immune response, and almost certainly untargeted generation of genomic uracil is harmful. In this chapter, we will discuss the mechanisms by which uracil in the genome can lead to cancer and influence disease progression. We will begin by outlining the molecular mechanisms by which dysfunctional uracil repair can lead to cancer. We will then discuss the role of enzymatic cytosine deamination in cancer. Next, we will outline how altering uracil incorporation into DNA by modulating the folate-mediated one-carbon pathway and thymidylate synthesis is both a source of genomic instability and an avenue for enhancing chemotherapy. Then, we will briefly discuss possible indirect mechanisms by which genomic uracil can attenuate genomic stability, e.g. through telomere maintenance. Finally, we will give an overview of genetic variants of genes involved in uracil homeostasis and their possible clinical significance.

7.1. Cancer Resulting from Dysfunctional Uracil-DNA glycosylases

There is strong evidence implicating inactive or dysfunctional uracil-DNA glycosylases in oncogenesis, but it is not clear to what extent this is caused by genomic uracil, alternative glycosylase substrates (e.g. thymine or oxidized pyrimidines) or so-called non-canonical functions of these enzymes, e.g. in epigenetics. In Chapter 4, we discussed the four uracil-DNA glycosylases — UNG, SMUG1, TDG, and MBD4 — that recognize and excise uracil from DNA as the first step in base excision repair (BER), and that account for the majority of uracil repair in mammalian cells. In addition, the mismatch repair

(MMR) pathway has been shown to process genomic uracil in certain contexts (*i.e.* within immunoglobulin gene loci during antibody affinity maturation). In the following, we will discuss the consequences of deficient or dysfunctional uracil repair in relation to cancer.

7.1.1. *UNG Deficiency and Mutagenesis via Enzymatic Cytosine Deamination*

UNG efficiently removes genomic uracil from "all" DNA contexts, but any persisting U:G mismatches are obviously highly mutagenic upon replication. It was therefore somewhat surprising that UNG-defective mice were viable, fertile, had only slight/moderate increase in mutation frequencies and no obvious early phenotype (Chapter 4). This raised the questions whether the mild effects were due to extensive backup repair activities, insufficient number of mutagenic U:G mismatches generated or too insensitive methods to detect aberrant phenotypes. However, UNG-knockout mice do develop B-cell lymphoma much more frequently than wild-type at 18 months[5,6] and this is associated with early-age clonal lymphoid hyperplasia in splenic B cells and possibly an immunological imbalance characterized by altered IFNγ, IL-6, and IL-2 levels.[7] These findings point towards mutagenic AID-mediated deamination during antibody affinity maturation as a likely oncogenic driver. UNG-knockout mouse cell lines and tissues and UNG-deficient human cells also have increased genomic uracil levels compared to wild-type.[8-10] Corroborating this, gene-specific mutation analysis in UNG-knockout mice showed 1.4- and 1.8-fold increases in the AID deamination targets *Bcl-6* and *Myc*, respectively.[11] Moreover, a high-AID-expressing lymph node tumour had 3-fold increased mutation levels in both *Bcl-6* and *c-Myc* loci, but not in the tumour suppressor gene *Tp53* that is not targeted by AID.[11] The C:G to T:A transition mutation frequency in UNG-knockout mice was increased as well, consistent with aberrant somatic hypermutation (SHM) caused by UNG-deficiency. Thus, UNG-deficiency may lead to B-cell neoplasia, and unrepaired uracils generated by AID outside of Ig loci in B cells is a likely contributor to oncogenesis. Recent work has also suggested that UNG protects telomeres from genomic uracil generated by mistargeted AID[5] (discussed in Section 7.4.1). Nevertheless, despite the evidence that UNG deficiency confers an oncogenic phenotype in mice, there is so far little evidence that it does so in humans. Indeed, only a few patients with inactivating mutations in UNG have been studied, and although they developed lymphoid hyperplasia, they had not developed cancer by the time of the study.[12] There are also differences in uracil repair between mouse and man, with humans relying more on UNG for uracil repair than mice, in which SMUG1 also plays a significant role.[13] Therefore, one could speculate that UNG deficiency or dysregulation may actually be a stronger threat to genomic integrity in humans than in mice.

Interestingly, UNG activity may also promote cancer development in some contexts. Plasma cell tumor development can be induced in mice by intraperitoneal mineral oil (pristane) injection and expedited with constitutive BAD (Bcl-xL) expression.[14] Using this model, it was shown that UNG-deficiency was associated with a considerable reduction and a delay in onset of plasma cell tumour development compared to UNG-proficient mice.[14] Interestingly, the canonical *IgH/Myc* translocations [t(12;15)s], usually associated with pristane-induced plasmacytomas, were completely absent in these UNG-deficient mice (versus a 14% incidence in UNG-proficient mice).[14] Notably, such translocations were also absent in AID-deficient mice.[15] Thus, restricting uracil processing by UNG showed a similar effect to restricting uracil generation by AID. A later study in mice harbouring dysregulated BCL6 expression demonstrated that UNG deficiency also prevents development of BCL6-driven diffuse large B-cell lymphoma (DLBCL).[16] Thus, UNG activity may be deleterious and lead to oncogenesis in some contexts, as in Bcl-xL-induced plasmacytomas and BCL6-driven DLBCL. This is probably connected to cancers that are driven by specific AID-dependent translocations (*IgH/Myc*). Here, processing by UNG may be important to generate strand breaks at the deamination sites to facilitate the translocation events, similarly to its role in CSR.

7.1.2. *Backup Uracil Repair by SMUG1 and Mismatch Repair*

Although historically considered a broad specificity backup for UNG2,[17,18] SMUG1's role in DNA repair is not redundant. Thus, SMUG1 depletion in UNG-knockout mouse embryonic fibroblasts (MEFs) increased C to T mutation frequencies.[19] Later experiments reported similar results upon complete SMUG1-knockout,[20] suggesting that SMUG1 indeed contributes to uracil repair *in vivo*. This was supported by the finding that about 50% of the U:G repair activity in mouse cell-free extracts could be attributed to SMUG1.[13] However, SMUG1 has little effect on CSR and genomic uracil levels in UNG-proficient cells.[21-23] Importantly, SMUG1 is active towards other lesions, e.g. 5-hydroxymethyluracil. Thus, the increase in mutation frequency observed upon SMUG1 depletion/knockout may have been a result of SMUG1 activity towards other lesions.

A prospective study on mice deficient in SMUG1, UNG, and DNA mismatch repair protein MSH2 showed that (1) SMUG1- and UNG/SMUG1-knockout mice both bred normally and remained healthy for at least 12 months; (2) MSH2- and UNG/MSH2-knockout mice showed a greatly decreased life expectancy, and (3) UNG/SMUG1/MSH2-knockout exhibited an even shorter lifespan and a greatly increased cancer predisposition (predominantly lymphoma).[20] A later study showed that although SMUG1-knockout did not mediate higher levels of uracil in DNA from a variety of tissues and UNG-knockout only mediated a slight increase, UNG/SMUG1-knockout mice exhibited a large increase in genomic uracil in all tissues examined (Fig. 7.1).[8] It was therefore concluded that SMUG1 serves as an able backup for UNG to keep genomic uracil at physiologically low levels and *vice versa*. In the same study, whole genome sequencing of tumors from >2-year-old UNG/SMUG1-knockout mice showed that the most abundant mutations were C to T transitions. These transitions were most frequent within CpG sequences. C to T transitions at CpGs are the

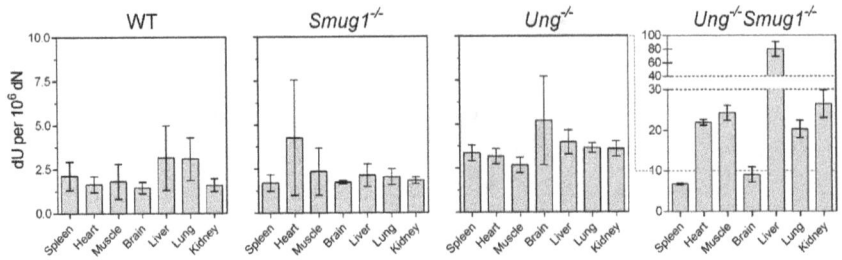

Fig. 7.1. UNG and SMUG1 are individually sufficient to maintain physiologically low genomic uracil levels (used with permission from[8]).

most frequent mutation observed in cancer and have primarily been attributed to deamination of 5-mC to generate T:G mispairs.[1] In the UNG/SMUG1-knockout mouse tumors there was, however, no bias towards mutation accumulation in regions with high density of CpG islands.[8] Nevertheless, the low rate of tumour formation in these mice may suggest that the main source of genomic uracil is non-mutagenic uracils opposite adenines. Furthermore, tumors in UNG/SMUG1/MSH2-knockout mice showed a greatly increased C to T transition mutation load. Since the MMR pathway can only repair genomic uracil opposite guanines, the results from these studies suggest that the MMR pathway repairs genomic uracil arising from deamination. Moreover, when both base excision and mismatch repair pathways are defective, the mutagenic effect of cytosine deamination is sufficient to increase cancer incidence without precluding mouse development. These experiments suggest that uracil processing involves both BER and MMR. Although either UNG or SMUG1 may individually be sufficient to maintain physiologically low levels of genomic uracil, its accumulation is not highly oncogenic unless the MMR pathway is also compromised.

7.1.3. *TDG and MBD4 are Mutated or Dysfunctional in some Cancers*

TDG and MBD4 are genuine DNA glycosylases that remove mismatched uracil, thymine and some oxidized pyrimidines from DNA.

They may have an important role in BER of U:G and T:G mismatches resulting from deamination of cytosine and 5-methylcytosine, respectively, particularly in CpG sequence contexts. However, TDG also has important roles in development and epigenetics, whereas MBD4 in addition to the glycosylase domain also carries a methyl-CpG-binding domain and interacts with mismatch repair protein MLH1. Although mutations in TDG and MDB4 are associated with cancer, this may well be independent of their roles in uracil repair.

TDG-deficiency confers embryonic lethality and epigenetic aberrations affecting expression of developmental genes.[24] 5-mC is the best understood and most studied epigenetic mark in DNA. It is essential to transcription regulation and aberrant methylation is clearly associated with cancer (reviewed in Ref. 25). Importantly, TDG is directly involved in demethylation by excising the oxidation products of 5-mC, (reviewed in Ref. 26). Moreover, it is well-established that 25% of all cancers are related to C to T transitions at CpG sites probably associated with 5-mC deamination, which results in the TDG substrate T:G.[27] The involvement of mutated TDG in cancer may well be via glycosylase-mediated aberrant demethylation or defective repair of T:G after 5-mC deamination, rather than genomic uracil repair. However, a role of TDG in uracil repair in the G2 to early G1 phase of the cell cycle, during which UNG expression is low, cannot be excluded.[28]

Several TDG polymorphisms (Section 7.5.2) have been reported in colorectal cancer (CRC) and it was speculated that TDG may play a role in CRC development.[29-31] It has also been shown that TDG is a coactivator of TP53, which in turn regulates *TDG* transcription.[32,33] In this sense, TDG can be classified as a tumour suppressor. Furthermore, the genoprotective effect of TDG via repair of deaminated 5-mC, was supported by the finding that a heterozygous *TDG* mutation in one patient resulted in protein loss, causing mutations in CpG islands that are normally methylated in colonic tissue.[31] TDG downregulation was also observed in gastric cancer, which also exhibited a reduction in expression of the methylcytosine dioxygenase TET1 and its product 5-hmC.[34] Thus, the oncogenic effect of TDG downregulation may be due to dysregulation of epigenetic mechanisms. Conversely, it has been shown that TDG levels were significantly higher in CRC tumour tissues than in normal tissues and that TDG upregulates WNT signalling.[35] The same study showed that knock-down of TDG inhibited proliferation of CRC cells both *in vitro* and *in vivo*,[35] indicating that TDG also functions as a cancer driver. Another recent study has shed light on a possible role of TDG in the regulation of histone acetylation.[36] The role of TDG in cancer development is therefore likely linked to its roles in DNA repair, epigenetic regulation as well as transcriptional activation and regulation of cell signalling.

MBD4 is likely important for mutation suppression, but it is unclear to what extent this is due to its uracil-DNA glycosylase activity. In contrast to the embryonic lethality of TDG-knockout mice, MBD4-knockout mice are fertile and develop normally. However, they exhibit a two- to three-fold increase in C:G to T:A transitions at CpG sites in the small intestines, indicating a possible role in repair of deaminated cytosines or thymine resulting from deaminated 5-mC at these sites.[37,38] Furthermore, when a cancer-susceptible mouse model heterozygous for the mutant adenomatous polyposis coli gene allele *Min* ($Apc^{+/min}$) was crossed with MBD4-knockout mice, they displayed an accelerated rate of tumour formation, with predominately C:G to T:A transitions at CpG sites.[37]

In humans, *MBD4* has been found to be mutated in 20–43% of mixed tissue samples from colorectal, endometrial, pancreatic and gastric cancers that exhibit microsatellite instability (MSI).[39-42] In microdissected samples from sporadic MSI colon tumours, truncated MBD4 mutations were found in as much as 89% of the samples.[42] Most of the mutations were located in an $(A)_{10}$-tract in exon 3 of *MBD4*, resulting in premature translation stop and a truncated protein lacking the glycosylase- and the MLH1-binding domains, but retaining the N-terminal methyl-binding domain. In by far the most cases, the frameshifts were heterozygous, but bi-allelic inactivation, either by a second frameshift or chromosomal loss of the second allele, has been observed in a few cases.[42,43] Notably, truncated MBD4 is still able to bind methylated CpGs, including those with T:G mismatches[44] and might thus exert a dominant negative effect by interfering with repair of such mismatches, and potentially also U:G mismatches. In support of this, truncated MBD4 was found to inhibit the activities of normal MBD4 and

bacterial Ung *in vitro* and also more than doubled the mutation frequency when expressed in MMR-deficient colon cancer cells.[45]
Interestingly, however, the mutation spectra revealed no bias towards C
to T transitions. This indicated that truncated MBD4 may also inhibit
the activity of other DNA glycosylases and potentially also other DNA
repair pathways. This was further supported by a study demonstrating
that truncated MBD4 predisposed to chromosomal rearrangements,
primarily reciprocal translocations and that this potentially was mediated via altered topoisomerase II function.[46] The above experiments
suggest that dysregulated uracil repair at best plays a minor role in the
mutagenic phenotype mediated by truncated MBD4.

7.2. Mutagenic U:G Generation by Enzymatic DNA-Cytosine Deamination

As discussed in Chapter 6, AID is central to B cell antibody maturation whereas several APOBEC3 paralogs are involved in innate
immune responses. Importantly, AID and APOBEC3A-H have been
shown to deaminate DNA cytosine *in vivo* (reviewed in Ref. 47).
Unfortunately, untargeted actions of these enzymes also hit several
genes that regulate cell proliferation.[48] If left unrepaired, the resulting
U:G mismatches may thus be fundamental causes of driver-mutations
in oncogenesis and tumour progression. Extensive whole-genome
and exome sequencing of human cancers have identified distinct
mutational signatures and their likely causes. Signatures of
AID/APOBEC are among the most common ones, but as could be
expected, several signatures are usually present in all cancers.[2,49]
Below, we discuss the emerging role of dysregulated action of DNA
deaminases in oncogenesis.

7.2.1. *AID-induced Mutations in B cell Cancers*

Upon antigen challenge, naïve mature B cells undergo somatic hypermutation (SHM) and class switch recombination (CSR). Both processes are dependent on AID-mediated deamination of cytosine to
uracil at *Ig* gene loci as described in Chapter 6. AID is required both

for mutations in variable region during SHM and for strand breaks in CSR. Thus, untargeted U:G mismatches generated by AID may be the substrate for point mutations in critical genes and for double strand breaks required for translocations. Recent studies have elucidated the importance of CSR and SHM dysregulation as an underlying cause of B-cell malignancies in particular. Indeed, aberrant SHM has been demonstrated in both human DLBCL and MALT lymphoma.[50,51] Knocking out AID results in significantly fewer mutations in several genes linked to B-cell tumourigenesis, while the uracil-DNA repair enzymes UNG and MSH2 protect against such AID-induced mutations.[48]

In lymphoma, specific DNA translocations are often diagnostic of specific subtypes. These translocations tend to involve *Ig* gene loci targeted by AID in normal SHM and CSR, and concomitant untargeted AID-mediated double-strand breaks at proto-oncogenes.[53] Fig. 7.2 shows an overview of AID-dependent translocations to *IgH* determined by inducing strand breaks at the *IgH* locus in mouse primary B cells overexpressing AID.[52] Examples of such *Ig*/proto-oncogene translocations are the prototypical *BCL1/IgH* translocation in mantle zone lymphoma, *BCL2/IgH* translocation in follicular lymphoma and *MYC/IgH* translocation in Burkitt's lymphoma.[54] Several of these proto-oncogenes are targeted by AID,[55,56] and AID is required for germinal center-derived lymphomagenesis.[57,58] AID-driven mutagenesis may also occur in later stages of B-cell malignancies. AID was shown to be highly expressed in several lymphoma patient samples and also correlated with ongoing somatic mutation.[59-63] Interestingly, lymphomas and leukemias show sub-clonal sequence heterogeneity, suggesting that AID may be a mutator during progression.[61,64-67] Still, to what extent this contributes to cancer progression and prognosis remains unclear. On one hand, AID expression was correlated with an unfavorable prognosis in chronic lymphocytic leukaemia/small lymphocytic lymphoma patients.[68] On the other hand, while AID was shown to be expressed in B-cell-like diffuse large-cell lymphomas, this was not correlated with intraclonal heterogeneity, suggesting that AID expression alone is not necessarily sufficient for ongoing SHM.[69] Interestingly, *Ig* gene translocations to

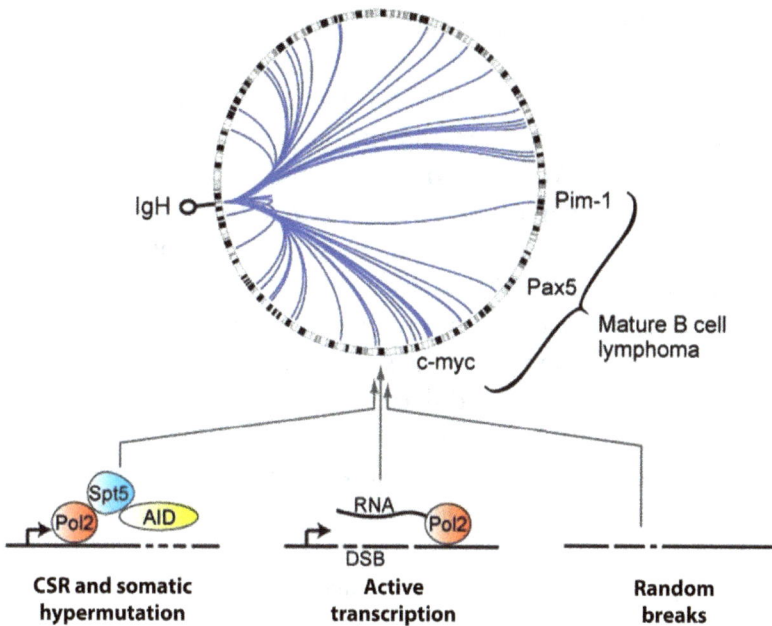

Fig. 7.2. AID-induced translocations involving IgH in B-cells overexpressing AID (from Ref. 52 with permission). Translocation-capture sequencing identifies genome-wide translocations (blue lines) to cleaved *IgH* in the mouse genome (circle). Active transcription increases translocations. AID-independent random breaks in AID$^{-/-}$ cells can be subtracted.

CCND1, *MYC*, and *RAS* genes among others can be found in 40% of multiple myeloma patients, suggesting a possible role for AID also in cancers arising from the late plasma cell differentiation stage of B cells.[70]

Studies in AID-knockout mice have demonstrated a strong reduction of translocations between *Myc* and the *IgH* variable region (Fig. 7.2).[15,52,71] Such translocations are likely directly mediated by AID since it preferentially deaminates single stranded DNA during transcription and since chromosomal breaks and rearrangements between *IgH* and *Myc* occur more frequently at transcription start sites.[72] Several lines of evidence also implicate active SHM in non-immunoglobulin genes including *BCL6, CD79, CD95* and *PAX,* [58,73-76] and a reduction of such mutations in the absence of AID.[48] Thus, AID is heavily implicated in the generation of oncogenic alteration of the genetic landscape and further underscores the need for strict regulation of AID to avoid off-target mutagenesis in B cells. AID has been shown to be expressed in several lymphoma subtypes[59,62,63,69,77] and cancer sequencing data have revealed AID mutational signatures in B-cell lymphoma and chronic lymphocytic leukaemia cells, particularly in regions of increased mutational accumulation, so-called *kataegis* regions.[2,62,78] (Fig. 7.4) Furthermore, we and others have demonstrated that AID expression is correlated with genomic uracil levels in lymphoma cell lines.[62,79] A role for uracil in lymphomagenesis is further corroborated by studies on the oncogenicity of UNG-knockout mice, as discussed in Section 7.1.1 in this chapter.

AID expression and activity have also been implicated in non-B-cell cancers and epithelial cancers, including T-cell lymphomas and gastric, liver, breast and ovarian cancers.[80-85] How the strict regulation of AID expression is overcome in these cancers is not clear. However, some studies indicate that aberrant AID expression in non-B cells can result from inflammation, bacterial and viral infections.[80,84,86-88]

7.2.2. A Role of AID/APOBECs in Cancer is Supported by Their Mutational Signatures

The APOBEC protein family consists of eleven members, AID, APOBEC1 (A1), APOBEC2 (A2), APOBEC3A-H (A3A-H), and APOBEC4 (A4), of which A1, AID and A3A-H exhibit *in vitro* cytosine and 5-mC deaminase activity (Table 7.1).[89-95] The two other members of the APOBEC-family, A2 and A4, have neither cytosine nor 5-mC deaminase activity.[92,96] The cytosine-deaminating APOBEC proteins are encoded by genes on chromosome 22 (A3A-H) and chromosome 12 (AID and A1), while the "non-deaminating" A2 and A4 are encoded by more distantly related genes on chromosomes 6 and 1, respectively (Table 7.1).[97] Furthermore, A1 has a known biological role in RNA editing.[47] Compared to the other APOBEC

Table 7.1. Human APOBEC-family Members[92,96,98,99]

Enzyme	Genomic Location	Exons/ Residues	Expression	DNA Seq Pref./ Hotspot	Physiological Target	*In vitro* Deaminase Activity C→U	5-mC→T
AID	12p13.31	5/198	Activated B-cells in spleen and lymph nodes	WR<u>C</u>Y	Ig genes	++	+
APOBEC1	12p13.31	6/223	Small intestine	T<u>C</u>W	ApoB mRNA	++	+
APOBEC2	6p21.1	3/224	Skeletal muscle, heart	Unknown	Unknown	–	–
APOBEC3A	22q13.1	5/199	Blood, keratinocytes, adipose, cervix, heart, lung, placenta, spleen, trachea	T<u>C</u>W	Virus DNA, retrotransp.	+++++	++++
APOBEC3B	22q13.1	8/382	Intestine, thyroid, cervix, mammary gland, lung, spleen, adipose	T<u>C</u>W	rVirus DNA, retrotransp.	+++	++
APOBEC3C	22q13.1	4/190	Blood, bladder, cervix, heart, lung, ovary, spleen	T<u>C</u>W	rVirus DNA, retrotransp.	+++	++
APOBEC3D	22q13.1	6/386	Blood, bladder, cervix, heart, lung, ovary, spleen, thymus	W<u>C</u>W	rVirus DNA, retrotransp.	++	ND
APOBEC3F	22q13.1	9/373	Adipose, bladder, cervix, colon, lung, ovary, spleen, thymus	T<u>C</u>W	rVirus DNA, retrotransp.	++	+
APOBEC3G	22q13.1	7/384	T-cells, blood, bladder, cervix, lung, ovary, spleen, thymus	C<u>C</u>C	rVirus DNA, retrotransp.	++	+
APOBEC3H	22q13.1	6/200	Thymus, blood, lung, colon, placenta, prostate, bladder	T<u>C</u>W	rVirus DNA, retrotransp.	++++	++++
APOBEC4	1q25.3	2/367	Testis	Unknown	Unknown	–	–

W = A or T, R = A or G, Y = C or T, rVirus = retrovirus.

family members, A3A and A3H have the highest *in vitro* deaminase activities on both C and 5-mC substrates.[92]

The characteristic features of APOBEC mutational signatures are the occurrence of simultaneous C to T transitions, C to G transversions and, less frequently, C to A transversions. These generally occur in TCA and TCT (TCW) sequence contexts.[2] Although these cytosine mutations all arise from deamination, they do not all end up as C to T transitions due to alternative downstream processing. C to T transitions are probably predominantly caused by cytosine deamination to uracil, and further replication over these U:G mismatches, in which U is recognized as a T by replicative polymerases if left unrepaired before replication. One known source of C to G transversions has been identified in yeast and mice, in which deaminated cytosines are removed by DNA glycosylases prior to replication, but not further processed. This leaves abasic sites that may be subject to translesion bypass replication by the deoxycytidyl transferase REV1 and DNA polymerase ζ (zeta).[100–103] Such abasic sites may also give rise to C to A transversions, by a yet unknown mechanism.[100]

The link between APOBECs other than AID and mutagenicity was first demonstrated by the mutator activity of A1, A3C, and A3G in *E. coli* and increased incidence of hepatocellular carcinomas in transgenic mice overexpressing A1.[90,104] The mutational spectra of A1 and A3G (as well as AID) showed an increase in C to T transition mutations, including mutations in central tumor suppressor genes such as *Tp53* and *Apc*.[105] It is now known that several APOBECs preferentially deaminate TCW sequence contexts and primarily result in C to T transitions and C to G transversions, depending on downstream processing.[2,106]

APOBEC mutational signatures are found in a larger variety of cancers and are more prevalent than AID mutational signatures.[2,104,106–120] Sequencing of exomes or complete genomes of more than 7,000 human cancers identified mutational signatures compatible with APOBEC activity in 16 out of 30 cancer subtypes.[2] Subsequent analyses including 10,952 exomes and 1,048 whole-genomes have extended these numbers to 22 out of 40 human cancer subtypes (Fig. 7.3). These landmark studies have revealed considerable mutational complexity and more than one signature was always identified

Fig. 7.3. Mutational signatures across human cancer show that enzymatic deamination is a source of mutagenesis in a wide range of cancer types. Mutational signatures attributed to enzymatic deamination by AID/APOBECs or downstream processing thereof are framed in red. Signatures 2 and 13 are attributed to APOBEC activity, and signature 9 is attributed to POLη processing downstream of AID. Figure obtained from the Catalogue of Somatic Mutations in Cancer.[49,125,126]

in each cancer subtype. Nevertheless, there are clearly identifiable patterns. Cancer subtypes with particularly high mutation loads were frequently dominated by APOBEC signatures unless there were known environmental mutational causes such as UV-irradiation and smoking. Interestingly, the mutational signature study by Alexandrov and colleagues considered C to T transitions at CpG dinucleotides, presumably from spontaneous 5-mC deamination, to be "aging signatures."[2] However, since A3 enzymes such as A3A can deaminate 5-mC, this signature may have contributions from APOBEC enzymatic deamination.[121–123]

Whereas APOBEC mutational signatures are prevalent in multiple cancer subtypes, AID signatures seem to be mainly enriched in regions of localised hypermutation in lymphomas and CLL (Fig. 7.4).[2,62,78] Such regions of localised hypermutations were dubbed *kataegis* regions (*kataegis* is ancient Greek for thunderstorm). *Kataegis* regions are also found in other cancer types and yeast models have confirmed that such regions most likely result from AID/APOBEC-mediated deamination.[3,124]

Enrichment of AID signatures in *kataegis* regions for B-cell tumors is in line with the model of AID-induced carcinogenesis in B cells, but does not exclude other contributing mechanisms. Other studies have e.g. confirmed A3B upregulation and corresponding mutation signatures in lymphoma and CLL, lung cancer, and ovarian cancers.[107,114,116,127–129]

APOBEC dysregulation in cancer

A3G has been implicated in both carcinogenesis and tumor survival. A3G was shown to be highly expressed in CRCs and suggested to contribute to hepatic metastasis.[130] In lymphoma, A3G seems to promote cancer cell survival and worsen clinical outcomes.[131–133] A3B was shown to be overexpressed in breast cancer, and this correlated with a doubling of C to T and overall base substitution mutations and an increase in genomic uracil levels, both of which were reversible by shRNA knockdown of A3B.[134] The same research group later showed that A3B was also overexpressed and correlated with its distinct

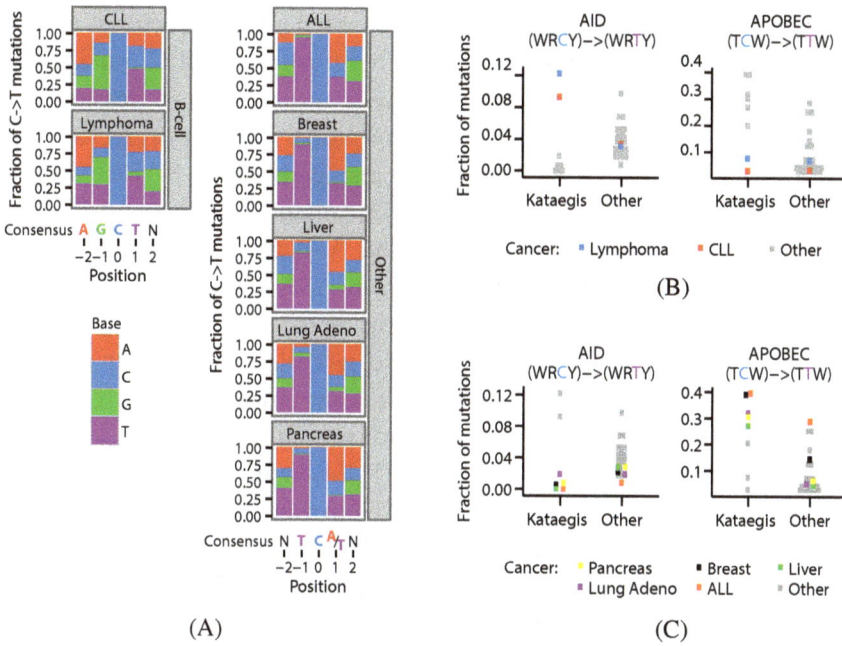

Fig. 7.4. (A) Differences in sequence contexts of C to T transition mutations in *kataegis* regions of lymphomas and CLL (AID mutational signature) versus ALL and cancers of the breast, liver, lung, and pancreas (APOBEC mutational signatures). Taken with permission from.[62] (B-C) Fractions of mutations occurring as C to T transitions at AID and APOBEC hotspot motifs in *kataegis* regions compared with other sites (data from Ref. 2). Mutations at AID motifs are enriched in *kataegis* regions of B-cell lymphoma and CLL, but not in other regions or cancers; mutations at APOBEC motifs are enriched in *kataegis* regions, except in B-cell lymphoma and CLL, and in other regions of some cancers. Cancers with *kataegis* regions from (A) are coloured separately in (B) and (C), with (B) showing B-cell lymphoma and CLL (B-cell in A) and (C) showing the other cancers. W = A or T, R = A or G, Y = C or T.

mutational signature in cancers of the bladder, cervix, lung, head/neck, and breast.[108] Others have shown similar findings of APOBEC mutational signatures in breast cancer.[2,106,109,112–114,118,135] Moreover, expression levels of A3B in established breast cancers has been positively correlated with both shorter disease-free survival and overall survival.[136] Indeed, APOBEC mutational signatures have been identified both as early and late events during tumor evolution in several cancers and seem to contribute considerably to later subclonal mutational events in lung cancer.[107,137,138] Therefore, APOBEC mutagenesis may be a driver of both carcinogenesis and cancer progression.

Intriguingly, a large fraction of the global population lacks A3B expression (0.9–6% of Africans and Europeans, 36.9–57.5% of East Asians and Amerindians, and 92.9% Oceanians).[139] This is due to a germline variant involving the A3A and A3B genes, which creates a fusion protein containing elements from A3A and B that interferes with A3B expression. Somewhat contrary to the abovementioned evidence, this variant confers increased risk of breast cancer and an increased number of APOBEC-associated mutational signatures.[140-144] This suggests that other APOBECs with mutational signatures similar to A3B, such as A1 or A3A/C/D/F/H, may also be significant mutational drivers in breast cancers.[96] Furthermore, a lack of A3B would possibly entail reduced viral defence capacity, increase in viral infections, and subsequent increased expression and off-target mutations by APOBECs other than A3B.[96,145,146] Nevertheless, A3-mediated off-target DNA mutagenesis does not seem to be an inevitable side effect of viral infections, since A3s are up-regulated following hepatitis B virus (HBV) infection without any subsequent increase in off-target A3-mediated mutations.[104] Still, HPV infections in breast epithelial cells as well as in head, neck and cervical cancer have been shown to induce anti-viral A3 expression and thus induce off-target A3 deamination-induced DNA mutations.[112,147-150] Moreover, expression of recombinant human A3A and A3B in yeast was found to induce genome-wide mutation signatures similar to that of breast cancers, suggesting a possible role for both A3A and A3B.[120] Furthermore, both A3B and A3A expression levels are correlated with *kataegic* mutation burden.[118,151] A possible role for A3A as a cancer driver is also strengthened by a study that implicated it as a significant mutational contributor in skin cancer in addition to UV-induced DNA damage.[152] Thus, A3B seems to be an important source of mutations in both breast cancer initiation and progression. However, the higher incidence of breast cancer in A3B-negative individuals highlights the complexity of this seemingly causal correlation.

Taken together, these findings suggest that enzymatic cytosine deamination by APOBEC enzymes may contribute to a significant fraction of mutations in a wide spectrum of human cancers. Still, although there is currently strong evidence for the contribution of several APOBECs in mutagenesis and carcinogenesis, more work is required to estimate and validate the relative carcinogenic contribution of each APOBEC enzyme for all relevant cancer subtypes, and subsequently further elucidate their possible diagnostic, prognostic and therapeutic roles.

7.3. Uracil Incorporation and Folate-mediated One-Carbon Metabolism

DNA uracil may arise by misincorporation of dUMP during DNA replication because replicative polymerases use dUTP and dTTP with equal efficiency (Section 3.1.4). Thus, the amount of uracil incorporation is determined by the dUTP/dTTP ratio in the cell. A critical determinant of this ratio is thymidylate synthase (TYMS) that converts dUMP to dTMP by employing 5,10-methylenetetrahydrofolate (5,10-CH_2THF, derived from folate) as methyl-donor. Chemotherapeutic agents that target thymidylate biosynthesis were among the first anti-cancer agents to be developed. These include antifolates against childhood acute leukemia developed by Sidney Farber in the 1940s[153] and 5-fluorouracil (5-FU) developed by Charles Heidelberger in the 1950s.[154] Fluoropyrimidines, such as 5-FU and capecitabine, and antifolates such as methotrexate and pemetrexed remain among the most efficient therapeutic agents in many hematologic malignancies and solid tumors, and their use is expanding to new cancer types and indications rather than declining.[155]

In this section, we will discuss to what extent uracil incorporation by modulation of the dUTP/dTTP pool affects oncogenesis. We will also discuss targeting pyrimidine biosynthesis, which has been a cancer treatment strategy for six decades.

7.3.1. *Is Uracil Misincorporation Caused by Modulation of dUTP and dTTP Pools Oncogenic?*

The two main enzymes that modulate the dUTP and dTTP pools are thymidylate synthase (TYMS, also known as TS) and deoxyuridine 5'-triphosphate nucleotidohydrolase (DUT, also known as dUTPase) (Fig. 3.1). TYMS contains binding pockets for both dUMP and its cofactor, the folate derivative 5,10-methylenetetrahydrofolate (5,10-CH$_2$THF).[156,157] Thus, folate is a necessary co-factor for TYMS activity, and folate deficiency can reduce 5,10-CH$_2$THF levels and TYMS activity, thereby increasing the dUTP/dTTP ratio and dUTP misincorporation during replication.[158,159] Nevertheless, it remains unclear to what extent increasing the genomic uracil load by modulating the dUTP/dTTP pool contributes to cancer risk. Cell culture and animal models have shown that folate deficiency leads to increased genomic uracil, DNA strand breaks, and genomic instability.[160–164] This would be consistent with a model in which high levels of genomic uracil induce futile repair cycles and increased strand breaks. Such a model is supported by studies in UNG-knockout mice, in which folate deficiency induced genomic uracil misincorporation without increasing DNA point mutations or tumor development.[165] Several studies, including epidemiological studies, furthermore indicate that low dietary folate may contribute to epithelial malignancies, especially CRC.[166–170] This picture is not entirely consistent, however, since some animal studies have demonstrated a protective effect of folate deficiency against experimental carcinogenesis and that folate supplementation has the opposite effect.[171] There is also growing evidence that higher circulating folate levels may contribute to prostate cancer progression.[172] The reason for these apparently conflicting results likely lies in the dual role of folate. In addition to being a precursor for 5,10-CH$_2$THF, it is also precursor for the universal methyl donor S-adenosylmethionine (SAM), which is necessary for a wide range of methylation reactions, including DNA, RNA and histone methylation. Aberrant levels of circulating folate may thus have adverse effects upon both epigenetic and epitranscriptomic regulation, and such effects are likely cell- and tissue dependent. Indeed, folate insufficiency has been associated with global DNA hypomethylation and may be reversed by folic acid supplementation in colorectal adenoma patients.[173]

In summary, it remains unclear to what extent folate-dependent modulation of the dUTP/dTTP ratio contributes to oncogenesis. The end result seems to be both tissue and context-specific, reliant on the degradation or disruption of other pathways, and/or marginal because (1) the biology of dietary folate and one-carbon metabolism is complex, (2) it is linked to the arguably more essential DNA methylation pathway, and (3) uracil repair may both aid and hamper mutagenesis.

7.3.2. *Fluoropyrimidines and Antifolates in Cancer Treatment — is Increased Genomic Uracil Involved in the Cytotoxic Effects?*

Several observations collectively led to the development of the first fluoropyrimidine used in cancer treatment. (1) When rats harboring acetylaminofluorene-induced hepatomas were administered [C^{14}]uracil, uracil was found to be incorporated in the nucleic acid fraction of the tumours, whereas this was not observed in the livers of healthy controls.[174] (2) Nucleic acid analogs, including synthetic uracil-analogs, had been found to exhibit anti-cancer effects.[175,176] (3) Fluorination was known to dramatically alter the biological properties of several aromatic compounds.[177] (4) Synthetic uracil analogs, including 6-azauracil, had been found to exhibit antitumor activity.[175] Based on these findings, Heidelberger and colleagues postulated that fluorinated pyrimidine analogues would disrupt DNA biosynthesis and kill cancer cells by targeting the pyrimidine biosynthesis pathway and in 1957 they synthesized the best-known fluoropyrimidine, 5-FU.[154] Thus, this represents one of the first examples of a rationally designed targeted therapy, although later research has demonstrated that the mechanism of action is more complicated than initially anticipated.

In 2004, approximately two million individuals received treatment with fluoropyrimidines[178] and more patients are treated today.

These compounds are primarily used in CRCs, but also in cancers of the pancreas, breast, head and neck region, stomach, and ovaries.[155] 5-FU is administered either intravenously or as an oral prodrug, for example tegafur-uracil, capecitabine, or S-1.[155] Capecitabine relies on the metabolic conversion of the prodrug to 5-FU in both liver and tumor tissue by enzymes such as carboxylesterase, cytidine deaminase, thymidine phosphorylase, and uridine phosphorylase.[179–181] This metabolic conversion not only makes it possible to avoid intravenous administration, but also to some degree ensures higher intra-tumor enzymatic conversion.[179,182,183] Tegafur-uracil and S-1 both contain tegafur, which is metabolised in the liver to 5′-hydroxytegafur and further to 5-FU. In addition, S-1 contains a 5-FU catabolism inhibitor (5-chloro-2,4-dihydroxypyridine), which inhibits dihydropyrimidine dehydrogenase (DPD), and potassium oxonate, which protects against gastrointestinal toxicity by inhibiting the enzyme orotate phosphoribosyltransferase (OPRT) that converts 5-FU to 5-fluorouridine monophosphate (5-FUMP).[184,185]

Antifolates are in extensive use for a wide range of cancers. The main antifolates in current clinical use are pemetrexed, raltitrexed, methotrexate, and pralatrexate.[186] The approved uses for these antifolates (including USA, Europe, Canada, Australia, and Japan) are: pemetrexed for non-small cell lung cancer (NSCLC) and mesothelioma; raltitrexed for CRC and mesothelioma; methotrexate for acute lymphoblastic leukemia (ALL), gestational trophoblastic disease, cutaneous T-cell lymphoma, non-Hodgkin lymphoma, osteosarcoma, head and neck, lung, and breast cancers; pralatrexate for refractory peripheral T-cell lymphoma.[186]

The common mechanism of action of both fluoropyrimidines and antifolates is inhibition of TYMS. Fluoropyrimidines inhibit TYMS by blocking its dUMP binding site, while antifolates inhibits TYMS by blocking the folate (5,10-CH$_2$THF) binding site. Both result in a reduction of dTTP and further disruption of nucleotide pools, which may result in blocked DNA replication and cell death. This phenomenon, named "thymineless death" was first observed in bacteria subjected to dTTP starvation, and later in mammalian cells.[187,188] Still, the exact mechanism underlying thymineless death remains elusive.[188]

Another overt effect of TYMS inhibition is the increase in dUMP levels. dUMP is converted to dUTP by the enzyme uridine monophosphate-cytidine monophosphate kinase (CMPK1). Presence of dUTP during replication carries the risk of misincorporation of dUTP instead of dTTP by DNA polymerases because replicative polymerases cannot distinguish well between the two nucleotides (Chapter 3).

dUTPase (DUT) maintains low intracellular dUTP levels by converting dUTP to dUMP. Several studies suggest that misincorporation of uracil into DNA and its prevention by DUT play a part in the anticancer effect of fluoropyrimidines and antifolates. Overexpression of *E. coli* dUTPase (dut) protects cells from fluorodeoxyuridine cytotoxicity, while depletion of DUT sensitises cells to fluorodeoxyuridine and pemetrexed.[189-191] Moreover, DUT expression levels inversely correlate with sensitivity to TYMS inhibitors, and high nuclear DUT levels are associated with a decrease in 5-FU efficacy, while DUT inhibitors have been shown to increase sensitivity to 5-FU in a xenograft model and increase 5-FU efficacy and genomic uracil levels in a CRC cell line.[192-195] Several cancers, including cancers that are treated with TYMS-inhibitors, such as CRC, non-small cell lung cancers and breast cancers, are found to overexpress DUT.[186] Recently, DUT inhibitors have been developed and the first human trials have been carried out.[194,196]

The anti-cancer mechanism of 5-FU and its derivatives is complex and is not limited TYMS inhibition. A major mode of action seems to be the conversion of 5-FU to several active metabolites directly and indirectly affecting RNA and DNA metabolism and function.[197] The conversion of 5-FU into 5-FdUMP affects DNA metabolism as described above by inhibiting TYMS and thereby disrupting *de novo* synthesis of dTMP from dUMP and subsequent unbalancing nucleotide pools. The resulting dUTP and 5-FdUTP can both be misincorporated into DNA and are potentially mutagenic and harmful through futile DNA excision repair.[198] The incorporated DNA uracil and 5-FU is recognised and excised from DNA mainly by the BER pathway.[199] 5-FU can also be converted to 5-fluorouridine triphosphate (5-FUTP) resulting in potential incorporation into RNA, which deleteriously affects the functions of ribosomal RNAs, transfer

RNAs, small nuclear RNAs, RNA exosomes, as well as inhibiting the functional conversion of uridine to pseudouridine in RNA.[199-205] Actually, more than 1,000-fold higher levels of 5-FU accumulate in RNA than in DNA in human cancer cell lines, suggesting that RNA-mediated effects contribute significantly to 5-FU cytotoxicity.[199] However, the relative contribution of each of these mechanisms remains unclear and the lack of mechanistic understanding of 5-FU is corroborated by the fact that only one-half of patients respond to 5-FU treatment.[197]

7.4. Indirect Contributions of Genomic Uracil to Cancer

7.4.1. *Dysregulated Telomeric Uracil Homeostasis Leads to Telomeric Instability*

Telomeres preserve chromosome stability and telomeric instability has been linked to cancer (reviewed in Ref. 206). Recent work indicates that uracil in telomeric DNA may be both a lesion that contributes to oncogenesis and a marker used to identify cells undergoing mutagenesis. In human B-lymphoblastoid cells, folate deficiency, which may increase genomic uracil, was shown to lengthen telomeres by 26% followed by rapid telomere attrition over 28 days. This was accompanied by increased chromosomal instability and mitotic dysfunction.[207] Furthermore, primary mouse UNG-knockout bone marrow cells have increased levels of telomeric uracil when compared with wild-type cells, apparently causing decreased binding of the telomere single-strand DNA-binding protein POT1, abnormal telomere lengthening and increased telomere fragility.[208] Thus, an increased genomic uracil load in telomeric DNA caused by increased dUTP misincorporation or UNG-deficiency leads to telomeric instability.

Importantly, AID may also target telomeric DNA. Indeed, telomeres and immunoglobulin switch (S) regions share several common features. They are downstream of RNA polymerase II promoters, contain C-rich template strands enriched in AID hotspot motifs, and form R-loops and non-coding transcripts that may form G-quadruplexes and recruit AID.[5,209-213] AID was shown to associate to telomeres with kinetics similar to those in the immunoglobulin Sμ region in CH12F3 cells, a murine B-cell line undergoing CSR.[5] When expressing the UNG inhibitor Ugi, the cells showed a four-fold increase in sister telomere loss (STL), a phenotype in which single chromatids lack a telomere. In agreement with this, stimulated UNG-deficient primary splenocytes showed an eight-fold increase in STL. However, telomere stability in CH12F3 was rescued when Ugi-expressing cells lacked AID, indicating that AID deaminates telomeres and that UNG protects them from this damage. To what extent AID-mediated deamination of telomeres contributes to human cancer remains, however, to be established.

7.4.2. *Processing of Uracil by Mismatch Repair (MMR) of DNA*

The post-replicative MMR factor MutSα, which is comprised of mutS homolog 2/6 (MSH2/6), can recognize and bind U:G mismatches generated by AID. This finding was somewhat unexpected since AID-induced mismatches arise in the G1 phase of the cell cycle, during which DNA lacks specific marks allowing strand selection. One would therefore expect U:G lesions in the G1-phase to be primarily repaired by the BER pathway. Although this is largely the case, disruption of either of the murine MMR genes *Msh2*, *Msh6*, *Mlh1*, *Pms2*, or *Exo1* results in altered SHM profiles as well as a two- to seven-fold reduction in CSR.[214-221] Furthermore, the mutational profile in activated B cells of mice deficient in both MSH2 and UNG shows that U:G mismatches are left unrepaired and converted to C to T and G to A mutations during replication, which represent the footprint of AID deamination.[222,223] A very similar effect is observed in humans, as patients lacking MSH6 or PMS2 exhibited impaired CSR.[224,225] Analysis of mutation frequencies in known AID target genes in B cells of *Ung−/−* and *Msh2−/−* single knockout mice indicates that both UNG2 and MSH2 contribute to error-free uracil repair in many of the genes while contributing to error-prone processing in others.[48] However, MMR was recently suggested to be a central mechanism to eliminate B cells at high risk of genomic instability caused by AID-induced U:G sites at the telomeres.[5] Nevertheless, in non-B cells it is doubtful that processing of genomic uracil by MMR plays an important role to prevent cancer, since MMR deficiency leads to microsatellite instability with severe clinical consequences regardless of uracil repair.

7.5. Germline Variants in Uracil-modulating Genes

The role of genomic uracil in oncogenesis makes the study of germline variants of several genes an interesting research topic. Traditionally, single nucleotide polymorphisms (SNPs) are gene variants that have a gene frequency higher than 1%, whereas less common ones are classified as mutations or rare variants. Due to negative selection, loss-of-function (LoF) variants are less frequent than neutral or near neutral variants. In diploid cells, LoF variants may remain undetected due to the presence of an active allele, as is the case in recessive disease. However, in some cases, sufficient activity requires the activity of both alleles and one LoF allele may give a phenotype (haplo-insufficiency). In other cases, allele variants may be dominant negative.

In this section, we will present a brief overview of the natural variation found within uracil-modulating genes and summarise some of the work done in this field. Indeed, several studies have associated variants in uracil-modulating genes with cancer, but it is probably correct to state that no definitive causal links have been established.

7.5.1. *Exome Variants in Uracil-modulating Genes*

The Exome Aggregation Consortium (ExAc) has used high throughput sequencing to identify protein-coding variants in a diverse population of 60,706 individuals.[226] With the caveat that the majority of the individuals are of European ancestry, this dataset provides a fairly unbiased view of naturally occurring protein-coding variants. The data show that genes essential in cultured cells have lower frequencies of frame-shifts or other protein-truncating LoF variants than do less critical genes, such as olfactory receptors and recessive disease genes.[226] Using the ExAc data, we asked whether different classes of uracil-modulating genes showed any differences in numbers or frequencies of different types of genetic variants. Specifically, we focused on genes involved in one-carbon metabolism and thymidylate synthesis, uracil-DNA glycosylases and other glycosylases, and the APOBEC family. Types of genetic variants ranged from those that alter codons without changing the amino acid (synonymous) to those that are potentially LoF in some (LoF_mixed) or all (LoF) of a gene's transcripts (Fig. 7.5).

All the genes we examined had hundreds of variants, but only a small fraction of these were potential LoF variants (Fig. 7.5A). The genes involved in one-carbon metabolism and thymidylate synthesis (*DUT* and *TYMS*, as well as dihydrofolate reductase, *DHFR*, and deoxythymidylate synthase kinase, *DTYMK*) had few missense and LoF variants. The few LoF variants observed generally had very low allele frequencies — particularly for *DHFR* and *TYMS* — and all of these were "LoF_mixed" variants with likely less severe effects in some of the genes' other transcripts (Fig 7.5B). Indeed, statistical models developed by ExAC suggest that *TYMS* is intolerant for both heterozygous and homozygous LoF variants (Fig. 7.5C). This is in line with TYMS being involved in several biologically important pathways, including maintaining low genomic uracil levels.

Of the four uracil-DNA glycosylases, *UNG* and *TDG* have the least frequent LoF variants and both genes are predicted to be intolerant of homozygous LoF variants (Fig. 7.5A-C). This observation may be related to the fact that both UNG and TDG have important roles in addition to DNA repair. UNG is essential in antibody affinity maturation, while TDG likely plays a major role in epigenetic regulation during embryonal development. Indeed, TDG knockout is embryonically lethal in mice.[24] In contrast, missense and LoF variants appear more frequent in other DNA glycosylases, and these genes appear to be tolerant of LoF variants.

(A)

(B)

(C)

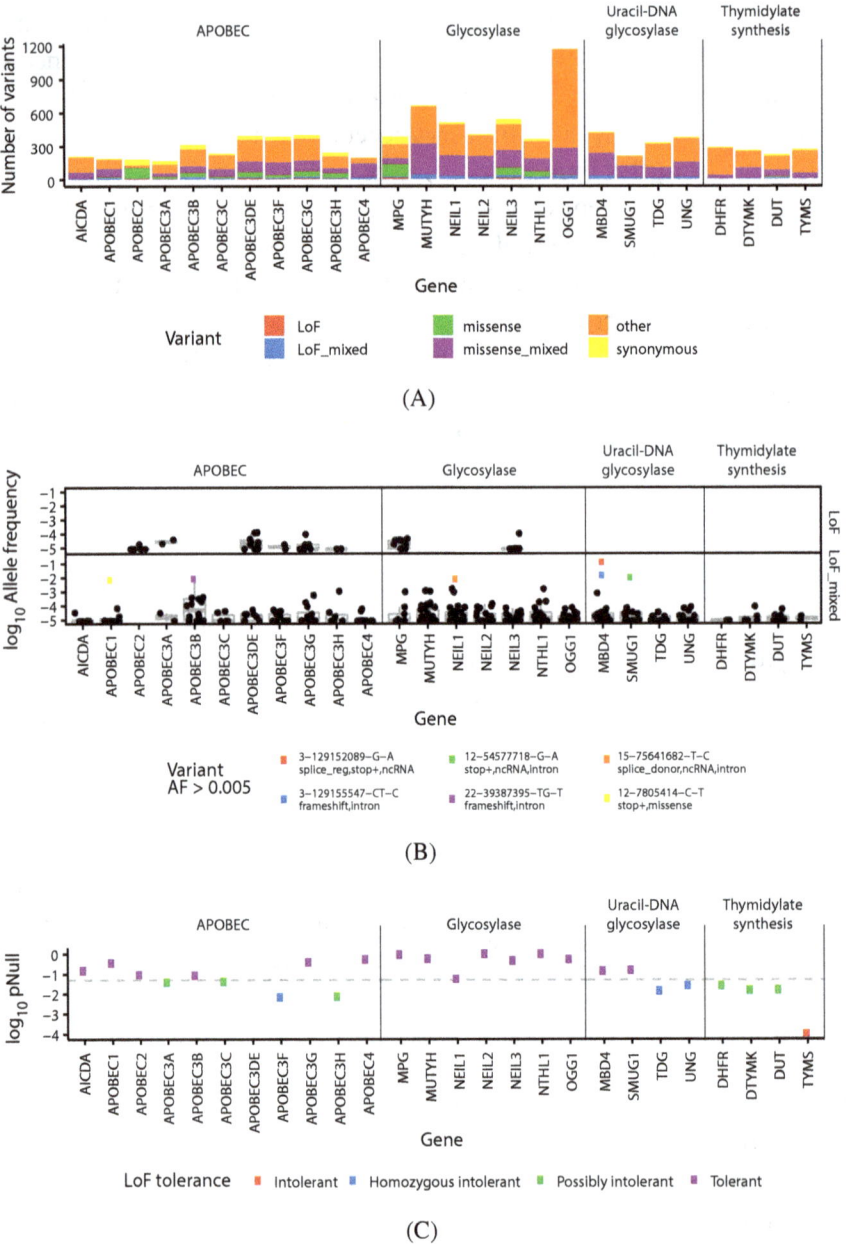

Fig. 7.5. (A) Number of unique variants observed for uracil-modulating genes and DNA glycosylases in the Exome Aggregation Consortium (ExAC) data (n = 60,706 individuals). "LoF" and "missense" are variants that are uniquely annotated as

Fig. 7.5. (*Continued*) loss-of-function and missense variants, respectively. "LoF_ mixed" and "missense_mixed" are variants that are annotated as LoF and missense variants in one of the gene's transcripts, respectively, and are annotated with a less severe functional consequence in other transcripts. (B) Allele frequencies for the LoF and LoF_mixed variants from (A). Variants with allele frequencies of > 0.5% are color coded and listed with their corresponding ExAc identification number. Variant 3-129152089-G-A (red color) is also listed in Table 7.3 as rs140696. (C) The probability of uracil-modifying genes being tolerant of both heterozygous and homozygous LoF variants (pNull). "Intolerant" genes have a high probability of being intolerant of both heterozygous and homozygous LoF variants (pLI > 0.9). "Homozygous intolerant" have a high probability of being intolerant of homozygous, but not heterozygous LoF variants (pRec > 0.9). "Possibly intolerant" are the remaining genes with pNull < 0.05 (grey, dashed horizontal line). Data for (A) and (B) were downloaded from the ExAC browser (http://exac.hms.harvard.edu); data for (C) are from Ref. 226 (Supplementary Table 13).

Notably, one *MBD4* variant (3-129152089) that causes a premature transcriptional stop codon was found to be relatively common (allele frequency of 0.1224). This variant is distinct from the mutation causing truncated, dominant negative MBD4 (described in Section 7.1.3) and lies within a transcript with a predicted short protein-coding reading frame that is different from MBD4. Moreover, the LoF was low-confidence, so it is unlikely to be a true LoF variant. Furthermore, genomic uracil in whole blood samples from individuals carrying this variant were measured and were not different from controls (Section 7.5.2).[227]

The genes in the APOBEC family show varying patterns in selection against LoF variants (Fig. 7.5). Unexpectedly, A1 was predicted to be tolerant to LoF variants, despite its necessity in lipid metabolism. Contrarily, A3F was predicted to be homozygous intolerant to LoF variants, whereas AID was tolerant according to the ExAc models. Although A3F is involved in the innate immune response against viral infection, AID is essential for adaptive immunity (discussed in Chapter 6) and its ablation leads to immunodeficiency. Furthermore, UNG is less tolerant to LoF variants, despite the fact that UNG deficiency leads to a less severe immunodeficiency than AID deficiency. One would therefore expect AID to be less tolerant than the other APOBECs involved in immunity, such as A3F. Although it is possible to interpret the data to conclude that either there is a heretofore undiscovered role of A3F that makes it more essential (and therefore less tolerant to LoF variants) than previously believed or that its role in immunity is important enough to be genetically conserved, this may just be an artefact of the analysis. Indeed, *AICDA* has the lowest number and the least frequent LoF variants of the APOBECs, both being lower than those of *UNG* and *TDG*.

In conclusion, analyzing genetic variants of uracil modulating genes affirms that some genes involved in maintaining uracil homeo-stasis are intolerant to LoF and therefore play a crucial biological role: However, as discussed in previous sections these roles are not limited to modulating genomic uracil levels.

7.5.2. *Other Variants in Uracil Modulating Genes and Their Proposed Association to Cancer*

Variants in the folate-mediated one-carbon synthesis pathway (Chapter 3.1 and Section 7.3.1) may modulate the cell's genomic uracil load, as well as the efficacy of fluoropyrimidines. For example, fluoropyrimidine treatment was reported to benefit from genotyping the *TYMS* gene.[228] Two *TYMS* polymorphisms have received particu-lar attention in oncogenesis: a 6 bp deletion in the 3′ UTR that influ-ences mRNA stability (1494del6) and a 28 bp variable number tandem repeat in the thymidylate synthase enhancer region (TSER) that contains an enhancer box and influences transcript levels. Several studies have shown an association between 1494del6 and TSER poly-morphisms and risk of cancer and disease progression in a wide range of malignancies.[229-232] However, a meta-analysis including 63 studies for 1494del6 (13,489 cases/16,297 controls) and 39 studies for TSER (19,707 cases/27,398 controls) failed to find an association when all studies were pooled together, but significant associations were observed among Asians after stratification with respect to cancer type and ethnicity.[233] Cancer-related SNPs in the *DUT* gene have also been explored, and reported to affect risk of persistent infection of human papillomavirus and further development to cervical cancer.[234] Another SNP was associated with increased genomic uracil in peripheral blood cells, but not with breast cancer risk.[227,235] In conclusion, variants in the *TYMS* and *DUT* genes may be associated with some cancer forms in some populations, but the underlying mechanisms do not necessarily involve genomic uracil.

APOBEC/AID expression levels have been strongly implicated in oncogenesis (Section 7.2), but comparatively few studies have addressed association between *APOBEC* germline variants and cancer (Table 7.2). Two studies have explored the role of germline AID gene (*AICDA*) variants in cancer risk. Jeon and colleagues studied seven *AICDA* SNPs and showed that four of them, all intronic, were associated with brain tumour risk in South Korean children.[236] One of the SNPs that was not associated with brain tumour risk (rs2028373) was also studied by another group in relation to *H. pylori*-related gastric cancer risk.[237] *H. pylori* infection is an established gastric cancer risk factor and has been shown to induce AID expression in gastric cells, and the SNP in question had been suggested to be functional in a previous study concerning AID and asthma.[238,239] However, the study revealed no association between *AICDA* SNP rs2028373 and gastric cancer.

In addition to *AICDA*, variants of the *APOBEC3* (*A3*) gene family have been studied in relation to cancer. Several groups have explored the anti-viral effects of *A3* SNPs that may have indirect oncogenic consequences (Table 7.2). Three variants upstream of *A3A* were associated with both breast and bladder cancer risk. Of these, one (rs1700526) was also associated with high A3B expression, as well as an APOBEC mutational signature in bladder cancer tumours.[240] Several other SNPs in *A3A* and *A3B* were associated with breast cancer risk or survival in one study while another study showed no association.[141,241] One variant entails a deletion between exon 5 of *A3A* and exon 8 of *A3B*, resulting in a fusion protein containing elements from both *A3A* and *A3B* (Section 7.2.2).[141] Although it was shown that breast cancer tumors from individuals expressing the fusion protein exhibited a more prominent APOBEC mutational signature than from individuals without, there are conflicting reports regarding its association with breast cancer risk.[141,142,240,242–244] Thus, A3 expression levels may play a more important role than their genetic variants.

Table 7.2.   Variants in *APOBEC* Genes Studied for Association with Cancer

Gene	SNP	dN	AA	MAF	Location	Disease Association	Ref.
AICDA	rs714629	C/G	—	0.4826°	~800 bp upstream	No association with childhood brain tumour risk.	236
	rs2580873	C/T	—	0.1510°	intron 1	Associated with brain tumour risk.	236
	rs12306110	C/G	—	0.1704°	intron 1	Associated with brain tumour risk.	236
	rs3794318	C/T	—	0.2105°	intron 2	Associated with brain tumour risk.	236
	rs2518144	A/G	—	0.4677§	intron 2	Associated with brain tumour risk.	236
	rs2028373	A/G	H155H	0.4096§	exon 4	No association with childhood brain tumour risk.	236
						No association with gastric cancer.	237
	rs11046349	G/T	—	0.2932°	~50 bp downstream	No association with childhood brain tumour risk.	236
A3A	rs1014971	A/G	—	0.4904°	>20 kbp upstream of A3A	Associated with breast and bladder cancer risk.	240
	rs1004748	C/G/T	—	0.4850°	>20 kbp upstream of A3A	Associated with breast and bladder cancer risk.	240
	rs17000526	A/G	—	0.4906°	>20 kbp upstream of A3A	Associated with breast and bladder cancer risk.	240
	rs5750715	A/T	—	0.2396°	~9 kbp upstream of A3A (eQTL)	No association with breast cancer risk or survival.	241
						Associated with breast cancer risk.	141
	rs5757402	C/T	—	0.3150°	~3 kbp upstream of A3A	Associated with low stage breast cancer.	241

Gene	SNP	Alleles	AA change	Frequency	Location	Association	Ref
	rs17370615	A/G	—	0.1672[a]	~3 kbp upstream of A3A	No association with breast cancer risk or survival.	241
						Associated with breast cancer risk.	243
	rs12157810	A/C	—	0.3340[a]	~300 bp upstream of A3A	No association with breast cancer risk or survival.	241
	rs12628403	A/C	—	0.0821[a]	intron 4	Associated with breast cancer risk.	141
A3B	rs2267398	C/T	—	0.2544[‡]	~900 bp upstream of A3B	No association with breast cancer risk or survival.	241
	rs8142462	C/T	—	0.1024[‡]	~600 bp upstream of A3B	No association with breast cancer risk or survival.	241
						Associated with breast cancer risk.	243
						Associated with increased breast cancer tumour size.	243
	rs28401571	A/C/T	—	0.3377[‡]	intron 1 (strong enhancer)	No association with breast cancer risk or survival.	241
						Associated with breast cancer risk.	243
	rs1065184	A/C/G/T	Y315*	0.4729[§]	exon 6	Associated with breast cancer risk.	243
	rs150925968	A/T	L489Q	0.0407[a]	exon 8	No association with breast cancer risk or survival.	241
	rs2142833	A/G	—	0.4423[a]	~4 kbp downstream of A3B (*eQTL*)	No association with breast cancer risk or survival.	241
	rs6001376	C/T	—	0.4000[a]	~15 kbp downstream of A3B (*eQTL*)	No association with breast cancer risk or survival.	241
						Associated with breast cancer risk.	243

(Continued)

Table 7.2.   (*Continued*)

Gene	SNP	dN	AA	MAF	Location	Disease Association	Ref.
A3A/B				0.009–0.923	deletion between exon 5 A3A and exon 8 A3B (results in fusion protein)	No association with breast cancer risk.	243
						Associated with breast cancer risk.	141
						Associated with breast cancer risk.	144
						No association with bladder cancer risk.	240
						Associated with breast cancer risk.	240
						No association with breast cancer risk.	242
						Associated with epithelial ovarian cancer risk.	244
A3G	rs7291971	C/G	—	0.4008[a]	~800 bp upstream of A3G	No association with hepatocellular carcinoma.	255
	rs5757463	C/G	—	0.0688[a]	~500 bp upstream of A3G	No association with hepatocellular carcinoma.	255
	rs5757465	C/T	F119F	0.2728[a]	exon 3	No association with hepatocellular carcinoma.	255
	rs8177832	A/G	H186R	0.0640[§]	exon 4	Associated with decreased risk for hepatocellular carcinoma.	255
	rs2011861	C/T	—	0.3828[a]	intron 5	Associated with increased risk for hepatocellular carcinoma.	255
A3H	rs139293	A/G/T	R18H/L	0.2685[§]	exon 2 (*eQTL*)	Associated with increased lung cancer risk.	256

MAF (Minor Allele Frequency) from [a]The Trans-Omics for Precision Medicine (TOPMed) Program, [§]ExAc, or [¥]The 1000 Genomes Project, *Premature stop, eQTL; Expression quantitative trait loci.

Several studies have reported SNPs in uracil-DNA glycosylase genes to be associated with elevated uracil levels or increased cancer predisposition (Table 7.3). Furthermore, SNPs in *TDG* and *SMUG1* have been associated with oesophageal squamous cell carcinoma, and several *SMUG1* SNPs modestly increase breast, bladder, and colon cancer risk.[235,245-248] Two SNPs in *SMUG1* and one in *UNG* were also reported to increase DNA uracil levels in whole blood.[227] Finally, several SNPs in *MBD4* have been associated with increased risk of oesophageal, lung, colon, and cervical cancers.[249-254] As outlined in Section 7.1.1 in this chapter, UNG and SMUG1 are the main glycosylases involved in genomic uracil repair, whereas the roles of MBD4 and TDG, although important, may be limited to U:G mismatches in CpG contexts. In addition, TDG has a defined role in epigenetic regulation. Dysregulated uracil levels resulting from UNG and SMUG1 may play a role in oncogenesis, though the interplay between uracil repair-related genomic instability and other oncogenic factors remains to be established.

7.6. Concluding Remarks

The fundamental cause of cancer is unrepaired DNA damage that cause cancer-driving mutations in critical genes. In a broad sense, such genes encode proteins involved in cell growth regulation and cellular communication. However, the path from DNA damage and mutation to clinical cancer is highly complex. Genomic uracil as a lesion has since the 1970s been known to occur as a consequence of spontaneous DNA-cytosine deamination and incorporation of dUMP residues during replication. If left unrepaired, the U:G mismatches from cytosine deamination will result in C to T transition mutations in 50% of the off-spring. In addition, errors during damage processing, both from U:G mismatches and U:A pairs, contribute to a wider mutational spectrum. All known organisms, as well as several viruses, have mechanisms that detect and correct genomic uracil, indicating that correction of this lesion is very important.

In the last decade, it has become clear that untargeted enzymatic deamination of genomic cytosine by members of the AID/APOBEC

Table 7.3.   Polymorphisms in Uracil DNA Glycosylase Genes Studied for Association with Cancer or Uracil Levels

Gene	SNP	dN	AA	MAF	Location	Functional Association	Ref.
UNG	rs34259	C/G	—	0.3037ᵃ	~2.5 kbp downstream	Increase in uracil levels in DNA (whole blood).	227
						No association with breask cancer risk.	235
	rs246079	A/G	—	0.3836ᵃ	*UNG* intron 6	Association with oesophageal cancer risk.	257
						Association with lung cancer risk.	258
	rs3219218	A/G	—	0.0162ᵃ	*UNG* intron 3	No association with oesophageal cancer risk.	257
	rs3890995	C/T	—	0.1713ᵃ	~2 kbp upstream	Association with bladder cancer disease progression.	259
	rs1018784	C/T	—	0.0660ᵃ	*UNG* intron 1	No increase in uracil levels in DNA (whole blood).	227
	rs3219266	C/T	—	0.0801ᵃ	*UNG* intron 6	No increase in uracil levels in DNA (whole blood).	227
	rs246085	C/T	—	0.1563ᵃ	~700 bp downstream	No increase in uracil levels in DNA (whole blood).	227
	rs3219235	–/T	—	0.2094$	*UNG* intron 4	Reported in >10 heterozygote cases of familial CRC.	29
SMUG1	rs2029166	C/T	—	0.3207ᵃ	~7 kbp upstream	Increase in uracil levels in DNA (whole blood).	227
						Association with oesophageal cancer risk.	245
						Association with breast cancer risk.	235

Gene	SNP	Allele	Protein	Frequency	Location	Effect	Ref
	rs7296239	C/T	—	0.4394°	~9 kbp upstream	Increase in uracil levels in DNA (whole blood).	227
						Association with breast cancer risk.	235
	rs2029167	A/G	—	0.4788°	~7 kbp upstream	Association with bladder cancer risk.	247
	rs2233921	G/T	—	0.2840°	~100 bp downstream	Association with CRC survival.	246
	rs1994356	A/G	—	0.2705°	~12 kbp downstream	No increase in uracil levels in DNA (whole blood).	227
	rs2233920	G/T	G15V	0.0077$	exon 3	Reported in 1 heterozygote case of familial CRC.	29
TDG	rs4135113	A/G/T	G199S	0.0465$	exon 5	Association with oesophageal cancer risk.	248
						No association with oesophageal cancer risk.	245
						No association with colon cancer risk.	260
						No association with colon cancer risk.	261
						No association with lung cancer risk.	262
						Genomic instability and more DSBs/cell transformation.	30
						Susceptibility to chromosomal damage.	263

(Continued)

Table 7.3. (*Continued*)

Gene	SNP	dN	AA	MAF	Location	Functional Association	Ref.
	rs4135054	C/T	—	0.2123[□]	intron 1	Association with oesophageal cancer risk.	245
	rs2700505	A/G	—	0.4481[□]	~14 kbp upstream	No increase in uracil levels in DNA (whole blood).	227
	rs4135150	C/T	—	0.0720[□]	~900 kbp downstream	Association with bladder cancer risk.	247
						Association with non-melanoma skin cancer risk.	264
						No increase in uracil levels in DNA (whole blood).	227
	rs2888805	C/G/T	V367M/L	0.10888[§]	exon 10	Association with non-melanoma skin cancer risk.	264
						Reported in >10 heterozygote cases of familial CRC.	29
						No association with lung cancer risk.	262
	rs167715	C/T	—	0.1080[□]	intron 2	Association with posthematopoetic cell transplant treatment-related mortality risk.	265
	rs2374327	A/T	—	0.2572[□]	intron 1	Association with posthematopoetic cell transplant treatment-related mortality risk.	265

rs703657	A/T	—	0.4483°	~9 kbp upstream	No increase in uracil levels in DNA (whole blood).	227
rs172814	A/G	—	0.1488§	~2 kbp upstream	No increase in uracil levels in DNA (whole blood).	227
rs4135063	C/T	—	0.2601°	intron 1	No increase in uracil levels in DNA (whole blood).	227
rs2723877	C/T	—	0.1080°	intron 1	No increase in uracil levels in DNA (whole blood).	227
rs1866074	C/T	—	0.3639°	intron 3	No increase in uracil levels in DNA (whole blood).	227
rs3829300	A/G	—	0.11114§	intron 2	Reported in 4 heterozygote cases of familial CRC.	29
rs3751209	A/G	—	0.2883§	intron 2	Reported in >10 heterozygote cases of familial CRC.	29
MBD4 rs2005618	C/T	—	0.2905°	intron 6	No association with oesophageal cancer risk.	266
rs10342	A/G/T	A273T/S	0.0811§	exon 3	No association with CRC risk.	267
					Loss of protein-DNA interaction (*in silico*).	268
rs3138373	A/G	—	0.1611°	~800 bp downstream	No association with oesophageal cancer risk.	266

(*Continued*)

Table 7.3. (*Continued*)

Gene	SNP	dN	AA	MAF	Location	Functional Association	Ref.
	rs140693	A/G	E346K	0.0510[$]	exon 3	Association with non-small cell lung cancer poor prognosis.	249
						Association with lung cancer risk.	252
						Association with lung cancer risk.	251
						Association with cervical cancer risk.	254
						Association with colon cancer risk.	253
	rs3138355	A/G	—	0.0847[o]	intron 5	Association with oesophageal cancer risk.	266
						No increase in uracil levels in DNA (whole blood).	227
	rs10470431	A/G	—	0.4250[o]	~9 kbp upstream	No increase in uracil levels in DNA (whole blood).	227
	rs140696	C/G/T	G471G	0.1224[$]	exon 6	No increase in uracil levels in DNA (whole blood).	227

MAF (Minor Allele Frequency) from [o]The Trans-Omics for Precision Medicine (TOPMed) Program or [$]ExAc.

family may give rise to mutational signatures associated with several important and common cancer forms. Thus, a small fraction of U:G mismatches escapes correction and causes distinct mutations that can be detected by high throughput sequencing. In this chapter, we have asserted that genomic uracil plays a role in cancer development and progression, and outlined the mechanisms by which this may occur. We have detailed how excessive genomic uracil generation and defective correction can contribute to cancer-driving mutations, but the role of genomic uracil in cancer is subtle and should only be thought of as one among several contributors to the oncogenic process. Dysfunctional uracil repair (Section 7.1) is unlikely to be a main cancer driver, probably because of the many layers of repair and tolerance of moderate increases in genomic uracil. As one example, UNG/SMUG1-deficient mice lack a general cancer phenotype, in spite of their increased genomic uracil burden. Furthermore, the redundancy involved in uracil repair makes it unlikely that a single dysfunctional gene can compromise the cell's ability to maintain tolerable levels of genomic uracil. Dietary folate and one-carbon metabolism (Section 7.3) may likewise influence genomic uracil levels, but the resulting uracil burden from these sources is apparently not primarily mutagenic. At present, sensitive methods to quantitate total genomic uracil are available. Finally, the role of genomic uracil in telomeres (Section 7.4.1) may be very relevant to cancer research, as it was shown that telomere stability can be disrupted by both uracil misincorporation and cytosine deamination.

The exception to the otherwise subtle role of genomic uracil in cancer is that of enzymatic deamination by AID/APOBECs. Mounting evidence accumulated during the past two decades strongly suggest that enzymatic deamination by AID contributes to B-cell lymphomagenesis. As discussed in Section 7.2.2, this is now a widely accepted phenomenon. The more recent work by the Catalogue of Somatic Mutations in Cancer team and others has brought into light that APOBECs contribute to mutagenesis in a wide variety of cancer types and are associated with the second most common mutational signature. This has established an irrefutable link between genomic uracil and cancer. However, the current question is to what extent enzymatic deamination by APOBECs and AID drives cancer. That is, is the mutational pressure from enzymatic deamination a major driver of cancer development and/or progression, or is the mutational pressure a factor that exacerbates or accelerates an already oncogenic phenotype resulting from the degradation of other anti-cancer safeguards? To answer these questions, novel methods to determine exact locations of genomic uracils are needed, as well as means to distinguish between uracil in U:A and U:G contexts. This will be outlined in more detail in Chapter 8.

References

1. Alexandrov, L. B., Stratton, M. R. Mutational signatures: the patterns of somatic mutations hidden in cancer genomes. *Curr Opin Genet Dev* **24**, 52–60 (2014).

2. Alexandrov, L. B., Nik-Zainal, S., Wedge, D. C., *et al.* Signatures of mutational processes in human cancer. *Nature* **500**, 415–421 (2013).

3. Lada, A. G., Dhar, A., Boissy, R. J., *et al.* AID/APOBEC cytosine deaminase induces genome-wide kataegis. *Biol Direct* **7**, 47; discussion 47 (2012).

4. Krokan, H. E., Sætrom, P., Aas, P. A., *et al.* Error-free versus mutagenic processing of genomic uracil — Relevance to cancer. *DNA Repair* **19**, 38–47 (2014).

5. Cortizas, E. M., Zahn, A., Safavi, S., *et al.* UNG protects B cells from AID-induced telomere loss. *J Exp Med* **213**, 2459–2472 (2016).

6. Nilsen, H., Stamp, G., Andersen, S., *et al.* Gene-targeted mice lacking the Ung uracil-DNA glycosylase develop B-cell lymphomas. *Oncogene* **22**, 5381–5386 (2003).

7. Andersen, S., Ericsson, M., Dai, H. Y., *et al.* Monoclonal B-cell hyperplasia and leukocyte imbalance precede development of B-cell malignancies in uracil-DNA glycosylase deficient mice. *DNA Repair (Amst)* **4**, 1432–1441 (2005).

8. Alsoe, L., Sarno, A., Carracedo, S., *et al.* Uracil Accumulation and Mutagenesis Dominated by Cytosine Deamination in CpG Dinucleotides in Mice Lacking UNG and SMUG1. *Sci Rep* **7**, 7199 (2017).

9. Galashevskaya, A., Sarno, A., Vagbo, C. B., *et al.* A robust, sensitive assay for genomic uracil determination by LC/MS/MS reveals lower levels than previously reported. *DNA Repair (Amst)* **12**, 699–706 (2013).

10. Kavli, B., Andersen, S., Otterlei, M., *et al.* B cells from hyper-IgM patients carrying UNG mutations lack ability to remove uracil from ssDNA and have elevated genomic uracil. *J Exp Med* **201**, 2011–2021 (2005).

11. Nilsen, H., An, Q., Lindahl, T. Mutation frequencies and AID activation state in B-cell lymphomas from Ung-deficient mice. *Oncogene* **24**, 3063–3066 (2005).

12. Imai, K., Slupphaug, G., Lee, W. I., *et al.* Human uracil-DNA glycosylase deficiency associated with profoundly impaired immunoglobulin class-switch recombination. *Nat Immunol* **4**, 1023–1028 (2003).

13. Doseth, B., Visnes, T., Wallenius, A., *et al.* Uracil-DNA glycosylase in base excision repair and adaptive immunity: species differences between man and mouse. *J Biol Chem* **286**, 16669–16680 (2011).

14. Kovalchuk, A. L., Ansarah-Sobrinho, C., Hakim, O., *et al.* Mouse model of endemic Burkitt translocations reveals the long-range boundaries of Ig-mediated oncogene deregulation. *Proc Natl Acad Sci U S A* **109**, 10972–10977 (2012).

15. Ramiro, A. R., Jankovic, M., Eisenreich, T., *et al.* AID is required for c-myc/IgH chromosome translocations in vivo. *Cell* **118**, 431–438 (2004).

16. Gu, X., Booth, C. J., Liu, Z., Strout, M. P. AID-associated DNA repair pathways regulate malignant transformation in a murine model of BCL6-driven diffuse large B-cell lymphoma. *Blood* **127**, 102–112 (2016).

17. Kavli, B., Sundheim, O., Akbari, M., *et al.* hUNG2 is the major repair enzyme for removal of uracil from U:A matches, U:G mismatches, and U in single-stranded DNA, with hSMUG1 as a broad specificity backup. *J Biol Chem* **277**, 39926–39936 (2002).

18. Nilsen, H., Haushalter, K. A., Robins, P., *et al.* Excision of deaminated cytosine from the vertebrate genome: role of the SMUG1 uracil-DNA glycosylase. *EMBO J* **20**, 4278–4286 (2001).

19. An, Q., Robins, P., Lindahl, T., Barnes, D. E. C → T mutagenesis and gamma-radiation sensitivity due to deficiency in the Smug1 and Ung DNA glycosylases. *EMBO J* **24**, 2205–2213 (2005).

20. Kemmerich, K., Dingler, F. A., Rada, C., Neuberger, M. S. Germline ablation of SMUG1 DNA glycosylase causes loss of 5-hydroxymethyluracil- and UNG-backup uracil-excision activities and increases cancer predisposition of Ung-/-Msh2-/- mice. *Nucleic Acids Res* **40**, 6016–6025 (2012).

21. Dingler, F. A., Kemmerich, K., Neuberger, M. S., Rada, C. Uracil excision by endogenous SMUG1 glycosylase promotes efficient Ig class switching and impacts on A:T substitutions during somatic mutation. *Eur J Immunol* **44**, 1925–1935 (2014).

22. Doseth, B., Ekre, C., Slupphaug, G., Krokan, H. E., Kavli, B. Strikingly different properties of uracil-DNA glycosylases UNG2 and SMUG1 may explain divergent roles in processing of genomic uracil. *DNA Repair (Amst)* **11**, 587–593 (2012).

23. Rada, C., Williams, G. T., Nilsen, H., *et al.* Immunoglobulin isotype switching is inhibited and somatic hypermutation perturbed in UNG-deficient mice. *Curr Biol* **12**, 1748–1755 (2002).

24. Cortazar, D., Kunz, C., Selfridge, J., *et al.* Embryonic lethal phenotype reveals a function of TDG in maintaining epigenetic stability. *Nature* **470**, 419–423 (2011).

25. Lakshminarasimhan, R., Liang, G. The Role of DNA Methylation in Cancer. *Adv Exp Med Biol* **945**, 151–172 (2016).

26. Kohli, R. M., Zhang, Y. TET enzymes, TDG and the dynamics of DNA demethylation. *Nature* **502**, 472–479 (2013).

27. Greenblatt, M. S., Bennett, W. P., Hollstein, M., Harris, C. C. Mutations in the p53 tumor suppressor gene: clues to cancer etiology and molecular pathogenesis. *Cancer Res* **54**, 4855–4878 (1994).

28. Hagen, L., Kavli, B., Sousa, M. M., *et al.* Cell cycle-specific UNG2 phosphorylations regulate protein turnover, activity and association with RPA. *EMBO J* **27**, 51–61 (2008).

29. Broderick, P., Bagratuni, T., Vijayakrishnan, J., *et al.* Evaluation of NTHL1, NEIL1, NEIL2, MPG, TDG, UNG and SMUG1 genes in familial colorectal cancer predisposition. *BMC Cancer* **6**, 243 (2006).

30. Sjolund, A., Nemec, A. A., Paquet, N., *et al.* A germline polymorphism of thymine DNA glycosylase induces genomic instability and cellular transformation. *PLoS Genet* **10**, e1004753 (2014).

31. Vasovcak, P., Krepelova, A., Menigatti, M., *et al.* Unique mutational profile associated with a loss of TDG expression in the rectal cancer of a patient with a constitutional PMS2 deficiency. *DNA Repair* **11**, 616–623 (2012).

32. da Costa, N. M., Hautefeuille, A., Cros, M.-P., *et al.* Transcriptional regulation of thymine DNA glycosylase (TDG) by the tumor suppressor protein p53. *Cell Cycle* **11**, 4570–4578 (2012).

33. Kim, E.-J., Um, S.-J. Thymine-DNA glycosylase interacts with and functions as a coactivator of p53 family proteins. *Biochem Biophys Res Commun* **377**, 838–842 (2008).

34. Yang, L., Yu, S.-J., Hong, Q., *et al.* Reduced Expression of TET1, TET2, TET3 and TDG mRNAs Are Associated with Poor Prognosis of Patients with Early Breast Cancer. *PLoS One* **10**, e0133896 (2015).

35. Xu, X., Yu, T., Shi, J., *et al.* Thymine DNA glycosylase is a positive regulator of Wnt signaling in colorectal cancer. *J Biol Chem* **289**, 8881–8890 (2014).

36. Henry, R. A., Mancuso, P., Kuo, Y.-M., *et al.* Interaction with the DNA Repair Protein Thymine DNA Glycosylase Regulates Histone Acetylation by p300. *Biochemistry* **55**, 6766–6775 (2016).

37. Millar, C. B., Guy, J., Sansom, O. J., *et al.* Enhanced CpG mutability and tumorigenesis in MBD4-deficient mice. *Science* **297**, 403–405 (2002).

38. Wong, E., Yang, K., Kuraguchi, M., *et al.* Mbd4 inactivation increases C→T transition mutations and promotes gastrointestinal tumor formation. *Proc Natl Acad Sci U S A* **99**, 14937–14942 (2002).

39. Bader, S., Walker, M., Hendrich, B., *et al.* Somatic frameshift mutations in the MBD4 gene of sporadic colon cancers with mismatch repair deficiency. *Oncogene* **18**, 8044–8047 (1999).

40. Riccio, A., Aaltonen, L. A., Godwin, A. K., *et al.* The DNA repair gene MBD4 (MED1) is mutated in human carcinomas with microsatellite instability. *Nat Genet* **23**, 266–268 (1999).

41. Yamada, T., Koyama, T., Ohwada, S., *et al.* Frameshift mutations in the MBD4/MED1 gene in primary gastric cancer with high-frequency microsatellite instability. *Cancer Lett* **181**, 115–120 (2002).

42. Bader, S., Walker, M., Harrison, D. Most microsatellite unstable sporadic colorectal carcinomas carry MBD4 mutations. *Br J Cancer* **83**, 1646–1649 (2000).

43. Bellacosa, A., Cicchillitti, L., Schepis, F., *et al.* MED1, a novel human methyl-CpG-binding endonuclease, interacts with DNA mismatch repair protein MLH1. *Proc Natl Acad Sci U S A* **96**, 3969–3974 (1999).

44. Hendrich, B., Hardeland, U., Ng, H. H., *et al.* The thymine glycosylase MBD4 can bind to the product of deamination at methylated CpG sites. *Nature* **401**, 301–304 (1999).

45. Bader, S. A., Walker, M., Harrison, D. J. A human cancer-associated truncation of MBD4 causes dominant negative impairment of DNA repair in colon cancer cells. *Br J Cancer* **96**, 660–666 (2007).

46. Abdel-Rahman, W. M., Knuutila, S., Peltomäki, P., *et al.* Truncation of MBD4 predisposes to reciprocal chromosomal translocations and alters the response to therapeutic agents in colon cancer cells. *DNA Repair* **7**, 321–328 (2008).

47. Knisbacher, B. A., Gerber, D., Levanon, E. Y. DNA Editing by APOBECs: A Genomic Preserver and Transformer. *Trends Genet* **32**, 16–28 (2016).

48. Liu, M., Duke, J. L., Richter, D. J., *et al.* G. Two levels of protection for the B cell genome during somatic hypermutation. *Nature* **451**, 841–845 (2008).

49. Forbes, S. A., Beare, D., Gunasekaran, P., *et al.* COSMIC: exploring the world's knowledge of somatic mutations in human cancer. *Nucleic Acids Res* **43**, D805–811 (2015).

50. Deutsch, A. J., Aigelsreiter, A., Staber, P. B., *et al.* MALT lymphoma and extranodal diffuse large B-cell lymphoma are targeted by aberrant somatic hypermutation. *Blood* **109**, 3500–3504 (2007).

51. Rossi, D., Berra, E., Cerri, M., *et al.* Aberrant somatic hypermutation in transformation of follicular lymphoma and chronic lymphocytic leukemia to diffuse large B-cell lymphoma. *Haematologica* **91**, 1405–1409 (2006).

52. Klein, Isaac A., Resch, W., Jankovic, M., *et al.* Translocation-Capture Sequencing Reveals the Extent and Nature of Chromosomal Rearrangements in B Lymphocytes. *Cell* **147**, 95–106 (2011).

53. Kuppers, R. Mechanisms of B-cell lymphoma pathogenesis. *Nat Rev Cancer* **5**, 251–262 (2005).

54. Kuppers, R., Dalla-Favera, R. Mechanisms of chromosomal translocations in B cell lymphomas. *Oncogene* **20**, 5580–5594 (2001).

55. Duquette, M. L., Pham, P., Goodman, M. F., Maizels, N. AID binds to transcription-induced structures in c-MYC that map to regions associated with translocation and hypermutation. *Oncogene* **24**, 5791–5798 (2005).

56. Duquette, M. L., Huber, M. D., Maizels, N. G-rich proto-oncogenes are targeted for genomic instability in B-cell lymphomas. *Cancer Res* **67**, 2586–2594 (2007).

57. Kotani, A., Kakazu, N., Tsuruyama, T., *et al.* Activation-induced cytidine deaminase (AID) promotes B cell lymphomagenesis in Emu-cmyc transgenic mice. *Proc Natl Acad Sci U S A* **104**, 1616–1620 (2007).

58. Pasqualucci, L., Bhagat, G., Jankovic, M., *et al.* AID is required for germinal center-derived lymphomagenesis. *Nat Genet* **40**, 108–112 (2008).

59. Greeve, J., Philipsen, A., Krause, K., *et al.* Expression of activation-induced cytidine deaminase in human B-cell non-Hodgkin lymphomas. *Blood* **101**, 3574–3580 (2003).

60. Hardianti, M. S., Tatsumi, E., Syampurnawati, M., *et al.* Expression of activation-induced cytidine deaminase (AID) in Burkitt lymphoma cells: rare AID-negative cell lines with the unmutated rearranged VH gene. *Leuk Lymphoma* **45**, 155–160 (2004).

61. Hardianti, M. S., Tatsumi, E., Syampurnawati, M., *et al.* Activation-induced cytidine deaminase expression in follicular lymphoma: association between AID expression and ongoing mutation in FL. *Leukemia* **18**, 826–831 (2004).

62. Pettersen, H. S., Galashevskaya, A., Doseth, B., *et al.* AID expression in B-cell lymphomas causes accumulation of genomic uracil and a distinct AID mutational signature. *DNA Repair (Amst)* **25**, 60–71 (2015).

63. Smit, L. A., Bende, R. J., Aten, J., *et al.* Expression of activation-induced cytidine deaminase is confined to B-cell non-Hodgkin's lymphomas of germinal-center phenotype. *Cancer Res* **63**, 3894–3898 (2003).

64. Campbell, P. J., Pleasance, E. D., Stephens, P. J., *et al.* Subclonal phylogenetic structures in cancer revealed by ultra-deep sequencing. *Proc Natl Acad Sci USA* **105**, 13081–13086 (2008).

65. Huemer, M., Rebhandl, S., Zaborsky, N., *et al.* AID induces intraclonal diversity and genomic damage in CD86(+) chronic lymphocytic leukemia cells. *Eur J Immunol* **44**, 3747–3757 (2014).

66. Ottensmeier, C. H., Thompsett, A. R., Zhu, D., *et al.* Analysis of VH genes in follicular and diffuse lymphoma shows ongoing somatic mutation and multiple isotype transcripts in early disease with changes during disease progression. *Blood* **91**, 4292–4299 (1998).

67. Spence, J. M., Abumoussa, A., Spence, J. P., Burack, W. R. Intraclonal diversity in follicular lymphoma analyzed by quantitative ultradeep sequencing of non-coding regions. *J Immunol* **193**, 4888–4894 (2014).

68. Leuenberger, M., Frigerio, S., Wild, P. J., *et al.* AID protein expression in chronic lymphocytic leukemia/small lymphocytic lymphoma is associated with poor prognosis and complex genetic alterations. *Mod Pathol* **23**, 177–186 (2010).

69. Lossos, I. S., Levy, R., Alizadeh, A. A. AID is expressed in germinal center B-cell-like and activated B-cell-like diffuse large-cell lymphomas and is not correlated with intraclonal heterogeneity. *Leukemia* **18**, 1775–1779 (2004).

70. Kuehl, W. M., Bergsagel, P. L. Early genetic events provide the basis for a clinical classification of multiple myeloma. *Hematology Am Soc Hematol Educ Program*, 346–352 (2005).

71. Dorsett, Y., Robbiani, D. F., Jankovic, M., *et al.* A role for AID in chromosome translocations between c-myc and the IgH variable region. *J Exp Med* **204**, 2225–2232 (2007).

72. Chiarle, R., Zhang, Y., Frock, R. L., *et al.* Genome-wide translocation sequencing reveals mechanisms of chromosome breaks and rearrangements in B cells. *Cell* **147**, 107–119 (2011).

73. Gordon, M. S., Kanegai, C. M., Doerr, J. R., Wall, R. Somatic hypermutation of the B cell receptor genes B29 (Igbeta, CD79b) and mb1 (Igalpha, CD79a). *Proc Natl Acad Sci U S A* **100**, 4126–4131 (2003).

74. Muschen, M., Re, D., Jungnickel, B., *et al.* Somatic mutation of the CD95 gene in human B cells as a side-effect of the germinal center reaction. *J Exp Med* **192**, 1833–1840 (2000).

75. Pasqualucci, L., Migliazza, A., Fracchiolla, N., *et al.* BCL-6 mutations in normal germinal center B cells: evidence of somatic hypermutation acting outside Ig loci. *Proc Natl Acad Sci U S A* **95**, 11816–11821 (1998).

76. Shen, H. M., Peters, A., Baron, B., *et al.* Mutation of BCL-6 gene in normal B cells by the process of somatic hypermutation of Ig genes. *Science* **280**, 1750–1752 (1998).

77. Pasqualucci, L., Guglielmino, R., Houldsworth, J., *et al.* Expression of the AID protein in normal and neoplastic B cells. *Blood* **104**, 3318–3325 (2004).

78. Kasar, S., Kim, J., Improgo, R., *et al.* Whole-genome sequencing reveals activation-induced cytidine deaminase signatures during indolent chronic lymphocytic leukaemia evolution. *Nat Commun* **6**, 8866 (2015).

79. Shalhout, S., Haddad, D., Sosin, A., *et al.* Genomic uracil homeostasis during normal B cell maturation and loss of this balance during B cell cancer development. *Mol Cell Biol* **34**, 4019–4032 (2014).

80. Endo, Y., Marusawa, H., Kinoshita, K., *et al.* Expression of activation-induced cytidine deaminase in human hepatocytes via NF-kappaB signaling. *Oncogene* **26**, 5587–5595 (2007).

81. Morisawa, T., Marusawa, H., Ueda, Y., *et al.* Organ-specific profiles of genetic changes in cancers caused by activation-induced cytidine deaminase expression. *Int J Cancer* **123**, 2735–2740 (2008).

82. Okazaki, I. M., Hiai, H., Kakazu, N., *et al.* Constitutive expression of AID leads to tumorigenesis. *J Exp Med* **197**, 1173–1181 (2003).

83. Pauklin, S., Sernandez, I. V., Bachmann, G., *et al.* Estrogen directly activates AID transcription and function. *J Exp Med* **206**, 99–111 (2009).

84. Shimizu, T., Marusawa, H., Matsumoto, Y., *et al.* Accumulation of somatic mutations in TP53 in gastric epithelium with Helicobacter pylori infection. *Gastroenterology* **147**, 407–417 e403 (2014).

85. Shinmura, K., Igarashi, H., Goto, M., *et al.* Aberrant expression and mutation-inducing activity of AID in human lung cancer. *Ann Surg Oncol* **18**, 2084–2092 (2011).

86. Mechtcheriakova, D., Svoboda, M., Meshcheryakova, A., Jensen-Jarolim, E. Activation-induced cytidine deaminase (AID) linking immunity, chronic inflammation, and cancer. *Cancer Immunol Immunother* **61**, 1591–1598 (2012).

87. Munoz, D. P., Lee, E. L., Takayama, S., *et al.* Activation-induced cytidine deaminase (AID) is necessary for the epithelial-mesenchymal transition in mammary epithelial cells. *Proc Natl Acad Sci U S A* **110**, E2977–2986 (2013).

88. Park, S. R. Activation-induced Cytidine Deaminase in B Cell Immunity and Cancers. *Immune Netw* **12**, 230–239 (2012).

89. Chelico, L., Pham, P., Calabrese, P., Goodman, M. F. APOBEC3G DNA deaminase acts processively 3′ → 5′ on single-stranded DNA. *Nat Struct Mol Biol* **13**, 392–399 (2006).

90. Harris, R. S., Petersen-Mahrt, S. K., Neuberger, M. S. RNA editing enzyme APOBEC1 and some of its homologs can act as DNA mutators. *Mol Cell* **10**, 1247–1253 (2002).

91. Hultquist, J. F., Lengyel, J. A., Refsland, E. W., *et al.* Human and rhesus APOBEC3D, APOBEC3F, APOBEC3G, and APOBEC3H demonstrate a conserved capacity to restrict Vif-deficient HIV-1. *J Virol* **85**, 11220–11234 (2011).

92. Ito, F., Fu, Y., Kao, S. A., *et al.* Family-Wide Comparative Analysis of Cytidine and Methylcytidine Deamination by Eleven Human APOBEC Proteins. *J Mol Biol* (2017).

93. Petersen-Mahrt, S. K., Neuberger, M. S. In vitro deamination of cytosine to uracil in single-stranded DNA by apolipoprotein B editing complex catalytic subunit 1 (APOBEC1). *J Biol Chem* **278**, 19583–19586 (2003).

94. Petersen-Mahrt, S. K., Harris, R. S., Neuberger, M. S. AID mutates E. coli suggesting a DNA deamination mechanism for antibody diversification. *Nature* **418**, 99–103 (2002).

95. Pham, P., Bransteitter, R., Petruska, J., Goodman, M. F. Processive AID-catalysed cytosine deamination on single-stranded DNA simulates somatic hypermutation. *Nature* **424**, 103–107 (2003).

96. Refsland, E. W., Harris, R. S. The APOBEC3 family of retroelement restriction factors. *Curr Top Microbiol Immunol* **371**, 1–27 (2013).

97. Swanton, C., McGranahan, N., Starrett, G. J., Harris, R. S. APOBEC Enzymes: Mutagenic Fuel for Cancer Evolution and Heterogeneity. *Cancer Discov* **5**, 704–712 (2015).

98. Harris, R. S., Liddament, M. T. Retroviral restriction by APOBEC proteins. *Nat Rev Immunol* **4**, 868–877 (2004).

99. Conticello, S. G. The AID/APOBEC family of nucleic acid mutators. *Genome Biol* **9**, 229 (2008).

100. Auerbach, P., Bennett, R. A., Bailey, E. A., *et al.* Mutagenic specificity of endogenously generated abasic sites in Saccharomyces cerevisiae chromosomal DNA. *Proc Natl Acad Sci U S A* **102**, 17711–17716 (2005).

101. Nelson, J. R., Lawrence, C. W., Hinkle, D. C. Deoxycytidyl transferase activity of yeast REV1 protein. *Nature* **382**, 729–731 (1996).

102. Jansen, J. G., Langerak, P., Tsaalbi-Shtylik, A., *et al.* Strand-biased defect in C/G transversions in hypermutating immunoglobulin genes in Rev1-deficient mice. *J Exp Med* **203**, 319–323 (2006).

103. Krijger, P. H. L., Tsaalbi-Shtylik, A., Wit, N., *et al.* Rev1 is essential in generating G to C transversions downstream of the Ung2 pathway but not the Msh2+Ung2 hybrid pathway. *Eur J Immunol* **43**, 2765–2770 (2013).

104. Henderson, S., Fenton, T. APOBEC3 genes: retroviral restriction factors to cancer drivers. *Trends Mol Med* **21**, 274–284 (2015).

105. Beale, R. C., Petersen-Mahrt, S. K., Watt, I. N., *et al.* Comparison of the differential context-dependence of DNA deamination by APOBEC enzymes: correlation with mutation spectra in vivo. *J Mol Biol* **337**, 585–596 (2004).

106. Nik-Zainal, S., Alexandrov, L. B., Wedge, D. C., *et al.* Mutational processes molding the genomes of 21 breast cancers. *Cell* **149**, 979–993 (2012).

107. de Bruin, E. C., McGranahan, N., Mitter, R., *et al.* Spatial and temporal diversity in genomic instability processes defines lung cancer evolution. *Science* **346**, 251–256 (2014).

108. Burns, M. B., Temiz, N. A., Harris, R. S. Evidence for APOBEC3B mutagenesis in multiple human cancers. *Nat Genet* **45**, 977–983 (2013).

109. Cancer Genome Atlas Research, N. Comprehensive molecular characterization of urothelial bladder carcinoma. *Nature* **507**, 315–322 (2014).

110. Davis, C. F., Ricketts, C. J., Wang, M., *et al.* The somatic genomic landscape of chromophobe renal cell carcinoma. *Cancer Cell* **26**, 319–330 (2014).

111. Harris, R. S. Molecular mechanism and clinical impact of APOBEC3B-catalyzed mutagenesis in breast cancer. *Breast Cancer Res* **17**, 8 (2015).

112. Henderson, S., Chakravarthy, A., Su, X., *et al.* APOBEC-mediated cytosine deamination links PIK3CA helical domain mutations to human papillomavirus-driven tumor development. *Cell Rep* **7**, 1833–1841 (2014).

113. Lawrence, M. S., Stojanov, P., Polak, P., *et al.* Mutational heterogeneity in cancer and the search for new cancer-associated genes. *Nature* **499**, 214–218 (2013).

114. Leonard, B., Hart, S. N., Burns, M. B., *et al.* APOBEC3B upregulation and genomic mutation patterns in serous ovarian carcinoma. *Cancer Res* (2013).

115. Nordentoft, I., Lamy, P., Birkenkamp-Demtroder, K., *et al.* Mutational context and diverse clonal development in early and late bladder cancer. *Cell Rep* **7**, 1649–1663 (2014).

116. Rebhandl, S., Huemer, M., Gassner, F. J., *et al.* APOBEC3 signature mutations in chronic lymphocytic leukemia. *Leukemia* **28**, 1929–1932 (2014).

117. Roberts, S. A., Gordenin, D. A. Hypermutation in human cancer genomes: footprints and mechanisms. *Nat Rev Cancer* **14**, 786–800 (2014).

118. Roberts, S. A., Lawrence, M. S., Klimczak, L. J., *et al.* An APOBEC cytidine deaminase mutagenesis pattern is widespread in human cancers. *Nat Genet* **45**, 970–976 (2013).

119. Saraconi, G., Severi, F., Sala, C., *et al.* The RNA editing enzyme APOBEC1 induces somatic mutations and a compatible mutational signature is present in esophageal adenocarcinomas. *Genome Biol* **15**, 417 (2014).
120. Taylor, B. J., Nik-Zainal, S., Wu, Y. L., *et al.* DNA deaminases induce break-associated mutation showers with implication of APOBEC3B and 3A in breast cancer kataegis. *Elife* **2**, e00534 (2013).
121. Carpenter, M. A., Li, M., Rathore, A., *et al.* Methylcytosine and normal cytosine deamination by the foreign DNA restriction enzyme APOBEC3A. *J Biol Chem* **287**, 34801–34808 (2012).
122. Suspene, R., Aynaud, M. M., Vartanian, J. P., Wain-Hobson, S. Efficient deamination of 5-methylcytidine and 5-substituted cytidine residues in DNA by human APOBEC3A cytidine deaminase. *PLoS One* **8**, e63461 (2013).
123. Wijesinghe, P., Bhagwat, A. S. Efficient deamination of 5-methylcytosines in DNA by human APOBEC3A, but not by AID or APOBEC3G. *Nucleic Acids Res* **40**, 9206–9217 (2012).
124. Roberts, S. A., Sterling, J., Thompson, C., *et al.* Clustered mutations in yeast and in human cancers can arise from damaged long single-strand DNA regions. *Mol Cell* **46**, 424–435 (2012).
125. Alexandrov, L. B., Nik-Zainal, S., Wedge, D. C., *et al.* Signatures of mutational processes in human cancer. *Nature* **500**, 415–421 (2013).
126. Forbes, S. A., Beare, D., Gunasekaran, P., *et al.* COSMIC: Catalogue of Somatic Mutations in Cancer — Home Page. *COSMIC: Catalogue of Somatic Mutations in Cancer — Home Page* (2017).
127. Sasaki, H., Suzuki, A., Tatematsu, T., *et al.* APOBEC3B gene overexpression in non-small-cell lung cancer. *Biomed Rep* **2**, 392–395 (2014).
128. Shinohara, M., Io, K., Shindo, K., *et al.* APOBEC3B can impair genomic stability by inducing base substitutions in genomic DNA in human cells. *Sci Rep* **2**, 806 (2012).
129. Waters, C. E., Saldivar, J. C., Amin, Z. A., *et al.* FHIT loss-induced DNA damage creates optimal APOBEC substrates: Insights into APOBEC-mediated mutagenesis. *Oncotarget* **6**, 3409–3419 (2015).
130. Ding, Q., Chang, C. J., Xie, X., *et al.* APOBEC3G promotes liver metastasis in an orthotopic mouse model of colorectal cancer and predicts human hepatic metastasis. *J Clin Invest* **121**, 4526–4536 (2011).
131. Jais, J. P., Haioun, C., Molina, T. J., *et al.* The expression of 16 genes related to the cell of origin and immune response predicts survival in elderly patients with diffuse large B-cell lymphoma treated with CHOP and rituximab. *Leukemia* **22**, 1917–1924 (2008).

132. Nowarski, R., Kotler, M. APOBEC3 cytidine deaminases in double-strand DNA break repair and cancer promotion. *Cancer Res* **73**, 3494–3498 (2013).

133. Nowarski, R., Wilner, O. I., Cheshin, O., *et al.* APOBEC3G enhances lymphoma cell radioresistance by promoting cytidine deaminase-dependent DNA repair. *Blood* **120**, 366–375 (2012).

134. Burns, M. B., Lackey, L., Carpenter, M. A., *et al.* APOBEC3B is an enzymatic source of mutation in breast cancer. *Nature* **494**, 366–370 (2013).

135. Nik-Zainal, S., Davies, H., Staaf, J., *et al.* Landscape of somatic mutations in 560 breast cancer whole-genome sequences. *Nature* **534**, 47–54 (2016).

136. Sieuwerts, A. M., Willis, S., Burns, M. B., *et al.* Elevated APOBEC3B correlates with poor outcomes for estrogen-receptor-positive breast cancers. *Horm Cancer* **5**, 405–413 (2014).

137. McGranahan, N., Favero, F., de Bruin, E. C., *et al.* Clonal status of actionable driver events and the timing of mutational processes in cancer evolution. *Sci Transl Med* **7**, 283ra254 (2015).

138. Zhang, J., Fujimoto, J., Zhang, J., *et al.* Intratumor heterogeneity in localized lung adenocarcinomas delineated by multiregion sequencing. *Science* **346**, 256–259 (2014).

139. Kidd, J. M., Newman, T. L., Tuzun, E., *et al.* Population stratification of a common APOBEC gene deletion polymorphism. *PLoS Genet* **3**, e63 (2007).

140. Komatsu, A., Nagasaki, K., Fujimori, M., *et al.* Identification of novel deletion polymorphisms in breast cancer. *Int J Oncol* **33**, 261–270 (2008).

141. Long, J., Delahanty, R. J., Li, G., *et al.* A common deletion in the APOBEC3 genes and breast cancer risk. *J Natl Cancer Inst* **105**, 573–579 (2013).

142. Nik-Zainal, S., Wedge, D. C., Alexandrov, L. B., *et al.* Association of a germline copy number polymorphism of APOBEC3A and APOBEC3B with burden of putative APOBEC-dependent mutations in breast cancer. *Nat Genet* **46**, 487–491 (2014).

143. Wellcome Trust Case Control, C., Craddock, N., Hurles, M. E., *et al.* Genome-wide association study of CNVs in 16,000 cases of eight common diseases and 3,000 shared controls. *Nature* **464**, 713–720 (2010).

144. Xuan, D., Li, G., Cai, Q., *et al.* APOBEC3 deletion polymorphism is associated with breast cancer risk among women of European ancestry. *Carcinogenesis* **34**, 2240–2243 (2013).

145. Koito, A., Ikeda, T. Intrinsic immunity against retrotransposons by APOBEC cytidine deaminases. *Front Microbiol* **4**, 28 (2013).

146. Malim, M. H., Bieniasz, P. D. HIV Restriction Factors and Mechanisms of Evasion. *Cold Spring Harb Perspect Med* **2**, a006940 (2012).

147. Ohba, K., Ichiyama, K., Yajima, M., *et al.* In vivo and in vitro studies suggest a possible involvement of HPV infection in the early stage of breast carcinogenesis via APOBEC3B induction. *PLoS One* **9**, e97787 (2014).

148. Ojesina, A. I., Lichtenstein, L., Freeman, S. S., *et al.* Landscape of genomic alterations in cervical carcinomas. *Nature* **506**, 371–375 (2014).

149. Vieira, V. C., Leonard, B., White, E. A., *et al.* Human papillomavirus E6 triggers upregulation of the antiviral and cancer genomic DNA deaminase APOBEC3B. *mBio* **5** (2014).

150. Warren, C. J., Xu, T., Guo, K., *et al.* APOBEC3A functions as a restriction factor of human papillomavirus. *J Virol* **89**, 688–702 (2015).

151. Jia, P., Pao, W., Zhao, Z. Patterns and processes of somatic mutations in nine major cancers. *BMC Med Genomics* **7**, 11 (2014).

152. Pham, P., Landolph, A., Mendez, C., *et al.* A biochemical analysis linking APOBEC3A to disparate HIV-1 restriction and skin cancer. *J Biol Chem* **288**, 29294–29304 (2013).

153. Farber, S., Diamond, L. K. Temporary remissions in acute leukemia in children produced by folic acid antagonist, 4-aminopteroyl-glutamic acid. *N Engl J Med* **238**, 787–793 (1948).

154. Heidelberger, C., Chaudhuri, N. K., Danneberg, P., *et al.* Fluorinated pyrimidines, a new class of tumour-inhibitory compounds. *Nature* **179**, 663–666 (1957).

155. Wilson, P. M., Danenberg, P. V., Johnston, P. G., *et al.* Standing the test of time: targeting thymidylate biosynthesis in cancer therapy. *Nat Rev Clin Oncol* **11**, 282–298 (2014).

156. Costi, M. P., Ferrari, S., Venturelli, A., *et al.* Thymidylate synthase structure, function and implication in drug discovery. *Curr Med Chem* **12**, 2241–2258 (2005).

157. Sommer, H., Santi, D. V. Purification and amino acid analysis of an active site peptide from thymidylate synthetase containing covalently bound 5-fluoro-2'-deoxyuridylate and methylenetetrahydrofolate. *Biochem Biophys Res Commun* **57**, 689–695 (1974).

158. Goulian, M., Bleile, B., Tseng, B. Y. Methotrexate-induced misincorporation of uracil into DNA. *Proc Natl Acad Sci USA* **77**, 1956–1960 (1980).

159. Hochster, H. The role of pemetrexed in the treatment of colorectal cancer. *Semin Oncol* **29**, 54–56 (2002).

160. Beetstra, S., Thomas, P., Salisbury, C., *et al.* Folic acid deficiency increases chromosomal instability, chromosome 21 aneuploidy and sensitivity to radiation-induced micronuclei. *Mutat Res* **578**, 317–326 (2005).

161. Duthie, S. J., Narayanan, S., Blum, S., *et al.* Folate deficiency in vitro induces uracil misincorporation and DNA hypomethylation and inhibits DNA excision repair in immortalized normal human colon epithelial cells. *Nutr Cancer* **37**, 245–251 (2000).

162. Duthie, S. J., Grant, G., Narayanan, S. Increased uracil misincorporation in lymphocytes from folate-deficient rats. *Br J Cancer* **83**, 1532–1537 (2000).

163. James, S. J., Yin, L. Diet-induced DNA damage and altered nucleotide metabolism in lymphocytes from methyl-donor-deficient rats. *Carcinogenesis* **10**, 1209–1214 (1989).

164. Pogribny, I. P., Muskhelishvili, L., Miller, B. J., James, S. J. Presence and consequence of uracil in preneoplastic DNA from folate/methyl-deficient rats. *Carcinogenesis* **18**, 2071–2076 (1997).

165. Linhart, H. G., Troen, A., Bell, G. W., *et al.* Folate deficiency induces genomic uracil misincorporation and hypomethylation but does not increase DNA point mutations. *Gastroenterology* **136**, 227–235.e223 (2009).

166. Duthie, S. J. Folate and cancer: how DNA damage, repair and methylation impact on colon carcinogenesis. *J Inherit Metab Dis* **34**, 101–109 (2011).

167. Krokan, H. E., Slupphaug, G., Kavli, B. in *The Base Excision Repair Pathway* 13–62 (WORLD SCIENTIFIC, 2015).

168. Sanjoaquin, M. A., Allen, N., Couto, E., *et al.* Folate intake and colorectal cancer risk: a meta-analytical approach. *Int J Cancer* **113**, 825–828 (2005).

169. Schernhammer, E. S., Ogino, S., Fuchs, C. S. Folate and vitamin B6 intake and risk of colon cancer in relation to p53 expression. *Gastroenterology* **135**, 770–780 (2008).

170. Choi, S. W., Mason, J. B. Folate and carcinogenesis: an integrated scheme. *J Nutr* **130**, 129–132 (2000).

171. Kim, Y. I. Folate, colorectal carcinogenesis, and DNA methylation: lessons from animal studies. *Environ Mol Mutagen* **44**, 10–25 (2004).

172. Rycyna, K. J., Bacich, D. J., O'Keefe, D. S. Opposing roles of folate in prostate cancer. *Urology* **82**, 1197–1203 (2013).

173. Pufulete, M., Al-Ghnaniem, R., Rennie, J. A., *et al.* Influence of folate status on genomic DNA methylation in colonic mucosa of subjects without colorectal adenoma or cancer. *Br J Cancer* **92**, 838–842 (2005).

174. Rutman, R. J., Cantarow, A., Paschkis, K. E. Studies in 2-acetylaminofluorene carcinogenesis. III. The utilization of uracil-2-C14 by preneoplastic rat liver and rat hepatoma. *Cancer Res* **14**, 119–123 (1954).

175. Jaffe, J. J., Handschumacher, R. E., Welch, A. D. Studies on the carcinostatic activity in mice of 6-azauracil riboside (azauridine), in comparison with that of 6-azauracil. *Yale J Biol Med* **30**, 168–175 (1957).

176. Stock, C. C. Experimental cancer chemotherapy. *Adv Cancer Res* **2**, 425–492 (1954).

177. Miller, J. A., Miller, E. C., Finger, G. C. On the enhancement of the carcinogenicity of 4-dimethylaminoazobenzene by fluoro-substitution. *Cancer Res* **13**, 93–97 (1953).

178. Ezzeldin, H., Diasio, R. Dihydropyrimidine dehydrogenase deficiency, a pharmacogenetic syndrome associated with potentially life-threatening toxicity following 5-fluorouracil administration. *Clin Colorectal Cancer* **4**, 181–189 (2004).

179. Liu, M., Cao, D., Russell, R., *et al.* Expression, characterization, and detection of human uridine phosphorylase and identification of variant uridine phosphorolytic activity in selected human tumors. *Cancer Res* **58**, 5418–5424 (1998).

180. Miwa, M., Ishikawa, T., Eda, H., *et al.* Comparative studies on the antitumor and immunosuppressive effects of the new fluorouracil derivative N4-trimethoxybenzoyl-5′-deoxy-5-fluorocytidine and its parent drug 5′-deoxy-5-fluorouridine. *Chem Pharm Bull (Tokyo)* **38**, 998–1003 (1990).

181. Mori, K., Hasegawa, M., Nishida, M., *et al.* Expression levels of thymidine phosphorylase and dihydropyrimidine dehydrogenase in various human tumor tissues. *Int J Oncol* **17**, 33–38 (2000).

182. Ishikawa, T., Utoh, M., Sawada, N., *et al.* Tumor selective delivery of 5-fluorouracil by capecitabine, a new oral fluoropyrimidine carbamate, in human cancer xenografts. *Biochem Pharmacol* **55**, 1091–1097 (1998).

183. Schuller, J., Cassidy, J., Dumont, E., *et al.* Preferential activation of capecitabine in tumor following oral administration to colorectal cancer patients. *Cancer Chemother Pharmacol* **45**, 291–297 (2000).

184. Saif, M. W., Syrigos, K. N., Katirtzoglou, N. A. S-1: a promising new oral fluoropyrimidine derivative. *Expert Opin Investig Drugs* **18**, 335–348 (2009).

185. Shirasaka, T., Shimamoto, Y., Fukushima, M. Inhibition by oxonic acid of gastrointestinal toxicity of 5-fluorouracil without loss of its antitumor activity in rats. *Cancer Res* **53**, 4004–4009 (1993).

186. Wilson, P. M., Fazzone, W., LaBonte, M. J., *et al.* Novel opportunities for thymidylate metabolism as a therapeutic target. *Mol Cancer Ther* **7**, 3029–3037 (2008).

187. Barner, H. D., Cohen, S. S. The induction of thymine synthesis by T2 infection of a thymine requiring mutant of Escherichia coli. *J Bacteriol* **68**, 80–88 (1954).

188. Guzmán, E. C., Martín, C. M. Thymineless death, at the origin. *Front Microbiol* **6**, 499 (2015).

189. Canman, C. E., Radany, E. H., Parsels, L. A., *et al.* Induction of resistance to fluorodeoxyuridine cytotoxicity and DNA damage in human tumor cells by

expression of Escherichia coli deoxyuridinetriphosphatase. *Cancer Res* **54**, 2296–2298 (1994).

190. Koehler, S. E., Ladner, R. D. Small interfering RNA-mediated suppression of dUTPase sensitizes cancer cell lines to thymidylate synthase inhibition. *Mol Pharmacol* **66**, 620–626 (2004).

191. Wilson, P. M., LaBonte, M. J., Lenz, *et al.* Inhibition of dUTPase induces synthetic lethality with thymidylate synthase-targeted therapies in non-small cell lung cancer. *Mol Cancer Ther* **11**, 616–628 (2012).

192. Hagenkort, A., Paulin, C. B. J., Desroses, M., *et al.* dUTPase inhibition augments replication defects of 5-Fluorouracil. *Oncotarget* **8**, 23713–23726 (2017).

193. Ladner, R. D., Lynch, F. J., Groshen, S., *et al.* dUTP nucleotidohydrolase isoform expression in normal and neoplastic tissues: association with survival and response to 5-fluorouracil in colorectal cancer. *Cancer Res* **60**, 3493–3503 (2000).

194. Miyahara, S., Miyakoshi, H., Yokogawa, T., *et al.* Discovery of a novel class of potent human deoxyuridine triphosphatase inhibitors remarkably enhancing the antitumor activity of thymidylate synthase inhibitors. *J Med Chem* **55**, 2970–2980 (2012).

195. Webley, S. D., Hardcastle, A., Ladner, R. D., *et al.* Deoxyuridine triphosphatase (dUTPase) expression and sensitivity to the thymidylate synthase (TS) inhibitor ZD9331. *Br J Cancer* **83**, 792–799 (2000).

196. Saito, K., Nagashima, H., Noguchi, K., *et al.* First-in-human, phase I dose-escalation study of single and multiple doses of a first-in-class enhancer of fluoropyrimidines, a dUTPase inhibitor (TAS-114) in healthy male volunteers. *Cancer Chemother Pharmacol* **73**, 577–583 (2014).

197. Longley, D. B., Harkin, D. P., Johnston, P. G. 5-fluorouracil: mechanisms of action and clinical strategies. *Nat Rev Cancer* **3**, 330–338 (2003).

198. van Laar, J. A., Rustum, Y. M., Ackland, S. P., *et al.* Comparison of 5-fluoro-2'-deoxyuridine with 5-fluorouracil and their role in the treatment of colorectal cancer. *Eur J Cancer* **34**, 296–306 (1998).

199. Pettersen, H. S., Visnes, T., Vagbo, C. B., *et al.* UNG-initiated base excision repair is the major repair route for 5-fluorouracil in DNA, but 5-fluorouracil cytotoxicity depends mainly on RNA incorporation. *Nucleic Acids Res* **39**, 8430–8444 (2011).

200. Ghoshal, K., Jacob, S. T. Specific inhibition of pre-ribosomal RNA processing in extracts from the lymphosarcoma cells treated with 5-fluorouracil. *Cancer Res* **54**, 632–636 (1994).

201. Gustavsson, M., Ronne, H. Evidence that tRNA modifying enzymes are important in vivo targets for 5-fluorouracil in yeast. *RNA* **14**, 666–674 (2008).

202. Patton, J. R. Ribonucleoprotein particle assembly and modification of U2 small nuclear RNA containing 5-fluorouridine. *Biochemistry* **32**, 8939–8944 (1993).

203. Samuelsson, T. Interactions of transfer RNA pseudouridine synthases with RNAs substituted with fluorouracil. *Nucleic Acids Res* **19**, 6139–6144 (1991).

204. Silverstein, R. A., Gonzalez de Valdivia, E., Visa, N. The incorporation of 5-fluorouracil into RNA affects the ribonucleolytic activity of the exosome subunit Rrp6. *Mol Cancer Res* **9**, 332–340 (2011).

205. Sun, X. X., Dai, M. S., Lu, H. 5-fluorouracil activation of p53 involves an MDM2-ribosomal protein interaction. *J Biol Chem* **282**, 8052–8059 (2007).

206. Artandi, S. E., DePinho, R. A. Telomeres and telomerase in cancer. *Carcinogenesis* **31**, 9–18 (2010).

207. Bull, C. F., Mayrhofer, G., O'Callaghan, N. J., *et al.* Folate deficiency induces dysfunctional long and short telomeres; both states are associated with hypomethylation and DNA damage in human WIL2-NS cells. *Cancer Prev Res (Phila)* **7**, 128–138 (2014).

208. Vallabhaneni, H., Zhou, F., Maul, R. W., *et al.* Defective repair of uracil causes telomere defects in mouse hematopoietic cells. *J Biol Chem* **290**, 5502–5511 (2015).

209. Balk, B., Maicher, A., Dees, M., *et al.* Telomeric RNA-DNA hybrids affect telomere-length dynamics and senescence. *Nat Struct Mol Biol* **20**, 1199–1205 (2013).

210. Pfeiffer, V., Crittin, J., Grolimund, L., Lingner, J. The THO complex component Thp2 counteracts telomeric R-loops and telomere shortening. *EMBO J* **32**, 2861–2871 (2013).

211. Schoeftner, S., Blasco, M. A. Developmentally regulated transcription of mammalian telomeres by DNA-dependent RNA polymerase II. *Nat Cell Biol* **10**, 228–236 (2008).

212. Storb, U. Why does somatic hypermutation by AID require transcription of its target genes? *Adv Immunol* **122**, 253–277 (2014).

213. Zheng, S., Vuong, B. Q., Vaidyanathan, B., *et al.* Non-coding RNA generated following lariat-debranching mediates targeting of AID to DNA. *Cell* **161**, 762–773 (2015).

214. Bardwell, P. D., Woo, C. J., Wei, K., *et al.* Altered somatic hypermutation and reduced class-switch recombination in exonuclease 1-mutant mice. *Nat Immunol* **5**, 224–229 (2004).

215. Ehrenstein, M. R., Neuberger, M. S. Deficiency in Msh2 affects the efficiency and local sequence specificity of immunoglobulin class-switch recombination: parallels with somatic hypermutation. *EMBO J* **18**, 3484–3490 (1999).

216. Li, A., Rue, M., Zhou, J., *et al.* Utilization of Ig heavy chain variable, diversity, and joining gene segments in children with B-lineage acute lymphoblastic leukemia: implications for the mechanisms of VDJ recombination and for pathogenesis. *Blood* **103**, 4602–4609 (2004).

217. Martomo, S. A., Yang, W. W., Gearhart, P. J. A Role for Msh6 But Not Msh3 in Somatic Hypermutation and Class Switch Recombination. *J Exp Med* **200**, 61–68 (2004).

218. Neuberger, M. S., Lanoue, A., Ehrenstein, M. R., *et al.* Antibody Diversification and Selection in the Mature B-cell Compartment. *Cold Spring Harbor Symp Quant Biol* **64**, 211–216 (1999).

219. Oers, J. M. M. v., Roa, S., Werling, U., *et al.* PMS2 endonuclease activity has distinct biological functions and is essential for genome maintenance. *Proc Natl Acad Sci U S A* **107**, 13384–13389 (2010).

220. Rada, C., Ehrenstein, M. R., Neuberger, M. S., Milstein, C. Hot spot focusing of somatic hypermutation in MSH2-deficient mice suggests two stages of mutational targeting. *Immunity* **9**, 135–141 (1998).

221. Schrader, C. E., Edelmann, W., Kucherlapati, R., Stavnezer, J. Reduced isotype switching in splenic B cells from mice deficient in mismatch repair enzymes. *J Exp Med* **190**, 323–330 (1999).

222. Rada, C., Di Noia, J. M., Neuberger, M. S. Mismatch recognition and uracil excision provide complementary paths to both Ig switching and the A/T-focused phase of somatic mutation. *Mol Cell* **16**, 163–171 (2004).

223. Xue, K., Rada, C., Neuberger, M. S. The in vivo pattern of AID targeting to immunoglobulin switch regions deduced from mutation spectra in msh2–/– ung-/- mice. *J Exp Med* **203**, 2085–2094 (2006).

224. Gardès, P., Forveille, M., Alyanakian, M.-A., *et al.* Human MSH6 deficiency is associated with impaired antibody maturation. *J Immunol* **188**, 2023–2029 (2012).

225. Péron, S., Metin, A., Gardès, P., *et al.* Human PMS2 deficiency is associated with impaired immunoglobulin class switch recombination. *J Exp Med* **205**, 2465–2472 (2008).

226. Lek, M., Karczewski, K. J., Minikel, E. V., *et al.* Analysis of protein-coding genetic variation in 60,706 humans. *Nature* **536**, 285–291 (2016).

227. Chanson, A., Parnell, L. D., Ciappio, E. D., *et al.* Polymorphisms in uracil-processing genes, but not one-carbon nutrients, are associated with altered

DNA uracil concentrations in an urban Puerto Rican population. *Am J Clin Nutr* **89**, 1927–1936 (2009).

228. Tan, B. R., Thomas, F., Myerson, R. J., *et al.* Thymidylate synthase genotype-directed neoadjuvant chemoradiation for patients with rectal adenocarcinoma. *J Clin Oncol* **29**, 875–883 (2011).

229. Mandola, M. V., Stoehlmacher, J., Muller-Weeks, S., *et al.* A novel single nucleotide polymorphism within the 5' tandem repeat polymorphism of the thymidylate synthase gene abolishes USF-1 binding and alters transcriptional activity. *Cancer Res* **63**, 2898–2904 (2003).

230. Mandola, M. V., Stoehlmacher, J., Zhang, W., *et al.* A 6 bp polymorphism in the thymidylate synthase gene causes message instability and is associated with decreased intratumoral TS mRNA levels. *Pharmacogenetics* **14**, 319–327 (2004).

231. Pullmann, R., Abdelmohsen, K., Lal, A., *et al.* Differential stability of thymidylate synthase 3'-untranslated region polymorphic variants regulated by AUF1. *J Biol Chem* **281**, 23456–23463 (2006).

232. Zhang, Z., Shi, Q., Sturgis, E. M., *et al.* Thymidylate synthase 5'- and 3'-untranslated region polymorphisms associated with risk and progression of squamous cell carcinoma of the head and neck. *Clin Cancer Res* **10**, 7903–7910 (2004).

233. Zhou, J.-Y., Shi, R., Yu, H.-L., *et al.* The association between two polymorphisms in the TS gene and risk of cancer: a systematic review and pooled analysis. *Int J Cancer* **131**, 2103–2116 (2012).

234. Wang, S. S., Gonzalez, P., Yu, K., *et al.* Common genetic variants and risk for HPV persistence and progression to cervical cancer. *PLoS One* **5**, e8667 (2010).

235. Marian, C., Tao, M., Mason, J. B., *et al.* Single nucleotide polymorphisms in uracil-processing genes, intake of one-carbon nutrients and breast cancer risk. *Eur J Clin Nutr* **65**, 683–689 (2011).

236. Jeon, S., Han, S., Lee, K., *et al.* Genetic variants of AICDA/CASP14 associated with childhood brain tumor. *Gen Mol Res* **12**, 2024–2031 (2013).

237. Hishida, A., Matsuo, K., Goto, Y., *et al.* No association between AICDA 7888 C/T polymorphism, Helicobacter pylori seropositivity, and the risk of atrophic gastritis and gastric cancer in Japanese. *Gastric Cancer* **13**, 43–49 (2010).

238. Matsumoto, Y., Marusawa, H., Kinoshita, K., *et al.* Helicobacter pylori infection triggers aberrant expression of activation-induced cytidine deaminase in gastric epithelium. *Nat Med* **13**, 470–476 (2007).

239. Noguchi, E., Shibasaki, M., Inudou, M., *et al.* Association between a new polymorphism in the activation-induced cytidine deaminase gene and atopic

asthma and the regulation of total serum IgE levels. *J Allergy Clin Immunol* **108**, 382–386 (2001).

240. Middlebrooks, C. D., Banday, A. R., Matsuda, K., *et al.* Association of germ-line variants in the APOBEC3 region with cancer risk and enrichment with APOBEC-signature mutations in tumors. *Nat Genet* **48**, 1330–1338 (2016).

241. Göhler, S., Da Silva Filho, M. I., Johansson, R., *et al.* Impact of functional germline variants and a deletion polymorphism in APOBEC3A and APOBEC3B on breast cancer risk and survival in a Swedish study population. *J Cancer Res Clin Oncol* **142**, 273–276 (2016).

242. Cescon, D. W., Haibe-Kains, B., Mak, T. W. APOBEC3B expression in breast cancer reflects cellular proliferation, while a deletion polymorphism is associated with immune activation. *Proc Natl Acad Sci U S A* **112**, 2841–2846 (2015).

243. Marouf, C., Göhler, S., Filho, M. I. D. S., *et al.* Analysis of functional germline variants in APOBEC3 and driver genes on breast cancer risk in Moroccan study population. *BMC Cancer* **16**, 165 (2016).

244. Qi, G., Xiong, H., Zhou, C. APOBEC3 deletion polymorphism is associated with epithelial ovarian cancer risk among Chinese women. *Tumour Biol* **35**, 5723–5726 (2014).

245. Li, W.-Q., Hu, N., Hyland, P. L., *et al.* Genetic variants in DNA repair pathway genes and risk of esophageal squamous cell carcinoma and gastric adenocarcinoma in a Chinese population. *Carcinogenesis* **34**, 1536–1542 (2013).

246. Pardini, B., Rosa, F., Barone, E., *et al.* Variation within 3'-UTRs of base excision repair genes and response to therapy in colorectal cancer patients: A potential modulation of microRNAs binding. *Clin Cancer Res* **19**, 6044–6056 (2013).

247. Xie, H., Gong, Y., Dai, J., Wu, X., Gu, J. Genetic variations in base excision repair pathway and risk of bladder cancer: A case-control study in the United States. *Mol Carcinog* **54**, 50–57 (2015).

248. Yang, X., Zhu, H., Qin, Q., *et al.* Genetic variants and risk of esophageal squamous cell carcinoma: a GWAS-based pathway analysis. *Gene* **556**, 149–152 (2015).

249. Dong, J., Hu, Z., Shu, Y., *et al.* Potentially functional polymorphisms in DNA repair genes and non-small-cell lung cancer survival: a pathway-based analysis. *Mol Carcinog* **51**, 546–552 (2012).

250. Hao, B., Wang, H., Zhou, K., *et al.* Identification of genetic variants in base excision repair pathway and their associations with risk of esophageal squamous cell carcinoma. *Cancer Res* **64**, 4378–4384 (2004).

251. Miao, R., Gu, H., Liu, H., *et al.* Tagging single nucleotide polymorphisms in MBD4 are associated with risk of lung cancer in a Chinese population. *Lung Cancer* **62**, 281–286 (2008).

252. Shin, M. C., Lee, S. J., Choi, J. E., *et al.* Glu346Lys polymorphism in the methyl-CpG binding domain 4 gene and the risk of primary lung cancer. *Jap J Clin Oncol* **36**, 483–488 (2006).

253. Song, J. H., Maeng, E. J., Cao, Z., *et al.* The Glu346Lys polymorphism and frameshift mutations of the Methyl-CpG Binding Domain 4 gene in gastrointestinal cancer. *Neoplasma* **56**, 343–347 (2009).

254. Xiong, X.-D., Luo, X.-P., Liu, X., *et al.* The MBD4 Glu346Lys polymorphism is associated with the risk of cervical cancer in a Chinese population. *Int J Gynecol Cancer* **22**, 1552–1556 (2012).

255. He, X.-T., Xu, H.-Q., Wang, X.-M., *et al.* Association between polymorphisms of the APOBEC3G gene and chronic hepatitis B viral infection and hepatitis B virus-related hepatocellular carcinoma. *World J Gastroenterol* **23**, 232–241 (2017).

256. Zhu, M., Wang, Y., Wang, C., *et al.* The eQTL-missense polymorphisms of APOBEC3H are associated with lung cancer risk in a Han Chinese population. *Sci Rep* **5**, 14969 (2015).

257. Yin, J., Sang, Y., Zheng, L., *et al.* Uracil-DNA glycosylase (UNG) rs246079 G/A polymorphism is associated with decreased risk of esophageal cancer in a Chinese population. *Med Oncol* **31**, 272 (2014).

258. Doherty, J. A., Sakoda, L. C., Loomis, M. M., *et al.* DNA repair genotype and lung cancer risk in the beta-carotene and retinol efficacy trial. *Int J Mol Epidemiol Genet* **4**, 11–34 (2013).

259. Wei, H., Kamat, A., Chen, M., *et al.* Association of polymorphisms in oxidative stress genes with clinical outcomes for bladder cancer treated with Bacillus Calmette-Guérin. *PLoS One* **7**, e38533 (2012).

260. Curtin, K., Ulrich, C. M., Samowitz, W. S., *et al.* Candidate pathway polymorphisms in one-carbon metabolism and risk of rectal tumor mutations. *Int J Mol Epidemiol Genet* **2**, 1–8 (2011).

261. Liu, A. Y., Scherer, D., Poole, E., *et al.* Gene-diet-interactions in folate-mediated one-carbon metabolism modify colon cancer risk. *Mol Nutr Food Res* **57**, 721–734 (2013).

262. Krześniak, M., Butkiewicz, D., Samojedny, A., *et al.* Polymorphisms in TDG and MGMT genes — epidemiological and functional study in lung cancer patients from Poland. *Ann Hum Genet* **68**, 300–312 (2004).

263. Wen-Bin, M., Wei, W., Yu-Lan, Q., Fang, J., Zhao-Lin, X. Micronucleus occurrence related to base excision repair gene polymorphisms in Chinese workers

occupationally exposed to vinyl chloride monomer. *J Occup Environ Med* **51**, 578–585 (2009).

264. Ruczinski, I., Jorgensen, T. J., Shugart, Y. Y., *et al.* A population-based study of DNA repair gene variants in relation to non-melanoma skin cancer as a marker of a cancer-prone phenotype. *Carcinogenesis* **33**, 1692–1698 (2012).

265. Thyagarajan, B., Lindgren, B., Basu, S., *et al.* Association between genetic variants in the base excision repair pathway and outcomes after hematopoietic cell transplantations. *Biol Blood Marrow Transplant* **16**, 1084–1089 (2010).

266. Yin, J., Shi, Y., Zuo, J., *et al.* Methyl-CpG binding domain 4 tagging polymorphisms and esophageal cancer risk in a Chinese population. *Eur J Cancer Prev* **24**, 100–105 (2015).

267. Tricarico, R., Cortellino, S., Riccio, A., *et al.* Involvement of MBD4 inactivation in mismatch repair-deficient tumorigenesis. *Oncotarget* (2015).

268. Allione, A., Guarrera, S., Russo, A., *et al.* Inter-individual variation in nucleotide excision repair pathway is modulated by non-synonymous polymorphisms in ERCC4 and MBD4 genes. *Mutat Res* **751–752**, 49–54 (2013).

Chapter 8

Quantification of Genomic Uracil

Antonio Sarno and Cathrine Broberg Vågbø*

Correct identification and quantification of uracil levels and location in DNA is essential to understand its biology. Uracil is generally present at very low levels in cellular DNA, so very sensitive analytical methods are necessary for its detection and quantification. Furthermore, small concentrations of contaminating molecules co-purifying with DNA may complicate analyses. In addition, naturally occurring modifications, as well as DNA damage occurring during sample preparation may be a challenge. One example of the latter is the strongly enhanced rate of deamination of cytosine to uracil in nucleotides and nucleosides compared with cytosine in double-stranded DNA. In spite of inherent problems, mass spectrometry has been used successfully for detection and quantification of uracil and other lesions or modifications in DNA. However, analyses of the sequence context in which such lesions occur, remains less well established.

8.1. Overview of Methods to Quantify Uracil in DNA

Several principally different methods have been employed to analyze genomic uracil (Fig. 8.1). In early work, the Lindahl laboratory

---

* *cathrine.b.vagbo@ntnu.no*

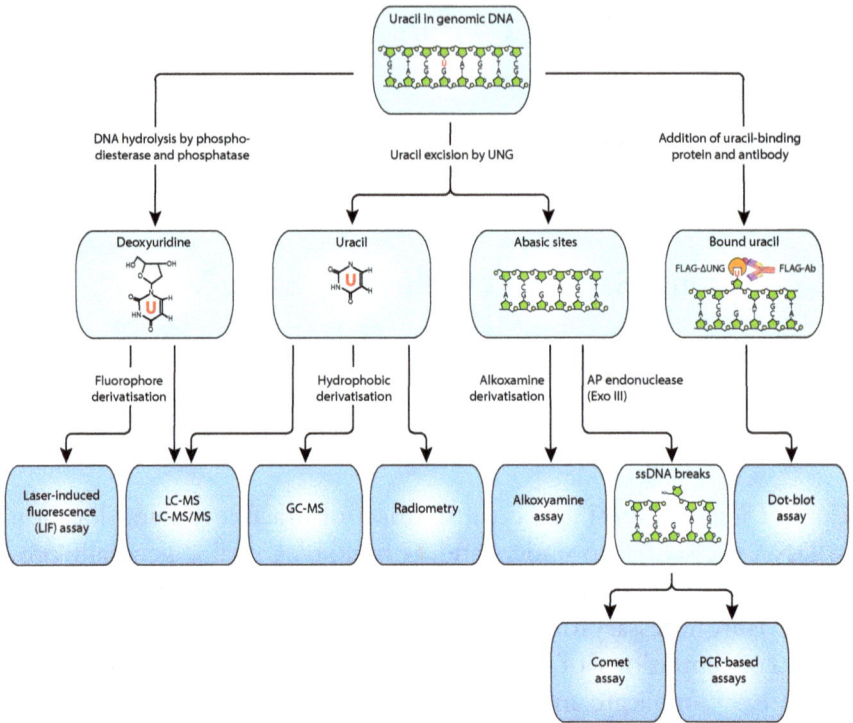

Fig. 8.1. Overview of methods used to quantify genome-wide uracil levels. Uracil itself may be quantified as enzymatically liberated nucleoside or base, or in full-length DNA using a uracil-binding protein and antibody. Uracil may also be quantified indirectly through the detection of UNG-induced abasic sites or single-strand breaks.

examined spontaneous deamination of cytosine to uracil at elevated temperature using DNA containing ^{14}C-cytosine. Following incubation of DNA for 0–8 days at 95°C, the DNA was enzymatically degraded to mononucleotides and analysed by paper chromatography.[1] These studies gave robust and valuable information on spontaneous generation of mutagenic U:G mismatches, but not about steady state levels of genomic uracil in cells. Several methods to quantify genomic uracil take advantage of the ability of UNG to specifically release uracil from DNA. Uracil is then quantified by mass spectrometry (MS), or the abasic site reacted with alkoxyamine and quantified by different immunochemical methods. Alternatively, the abasic sites are cleaved and quantified by the comet assay or by ligation-mediated PCR and other PCR methods. Importantly, the latter may in principle be used to identify the sequence context of genomic uracil. Uracil in DNA has also been quantified using uracil-binding proteins in combination with immune-detection. This method should also have the potential of isolating and sequencing uracil-containing DNA.

In this chapter, we will first give an overview of MS-based methods to determine genomic uracil. Subsequently we will describe several non-MS methods that are currently being used or used previously.

8.2. Quantification of Genomic Uracil by Gas-Chromatography-Mass Spectrometry (GC-MS) and Liquid Chromatography–Mass Spectrometry (LC-MS)

Analysis by GC-MS and LC-MS combines the advantages of high selectivity and high specificity. Genomic uracil can be quantified either after release of uracil by UNG or as deoxyuridine after enzymatic degradation of DNA to the constituent deoxyribonucleosides. Quantification of released uracil is usually by GC-MS after derivatisation of uracil (Fig. 8.1). Quantification of deoxyuridine is usually by LC-MS/MS. MS is arguably the approach with the highest specificity and sensitivity. Indeed, MS-based methods can distinguish between a vast number of different chemical DNA modifications, and are sensitive enough to detect very low DNA uracil levels (e.g. 1 uracil per 10^7 normal nucleosides).[2–8] MS relies on the ionization of analyte molecules and their subsequent separation based on mass-to-charge (m/z) ratios. Additional specificity may be achieved by the use of tandem MS (MS/MS) to measure both intact ions and fragment ions produced upon gas-phase collision of the molecules, thereby increasing the confidence in their identity (Fig. 8.2).

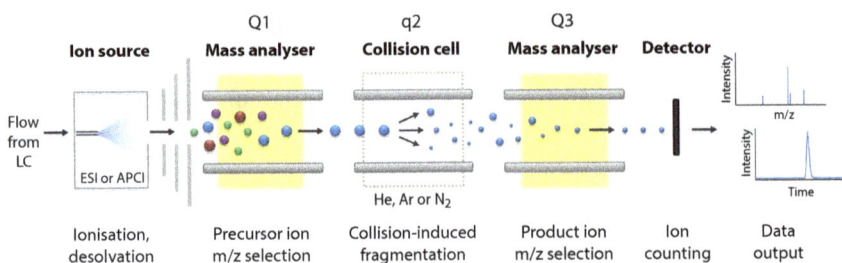

Fig. 8.2. Schematic representation of a triple quadrupole mass spectrometer. The instrument converts the sample into gaseous ions and analyses the ions based on their mass-to-charge (m/z) ratio and fragmentation pattern to determine their identity and quantity. The first and third quadrupoles (Q1 and Q3) act as mass filters enabling only ions of selected m/z to pass, and the second quadrupole (q2) acts as a collision cell in which the ions fragment due to collisions with an inert gas. In targeted mode, Q1 sequentially selects individual ions (precursor ions) for fragmentation in q2, whereupon Q3 selects one or several of the product ions produced. Ions reaching the detector are counted and converted into an electric signal. LC; liquid chromatography, ESI; electrospray ionisation, APCI; atmospheric pressure chemical ionization, Q; quadrupole.

To cope with the extreme complexity of biological samples, it is often necessary to separate the target molecule from the bulk of the thousands of other different molecules in the sample before presenting it to the mass spectrometer. Thus, chromatographic separation is generally employed prior to MS. Both LC and GC enable such a separation. The degree of analyte retention in LC depends on the physio-chemical properties of the molecule in solution, and the LC retention time thus offers an additional characteristic that may be used for molecule identification. These several layers of specificity — LC retention, mass-to-charge ratios, and molecular fragmentation pattern — make the combination LC-MS/MS a powerful analytical tool. When MS is coupled to GC, molecules are separates according to their volatility, also providing significant added specificity to the analysis compared to MS alone.[5,6,9,10]

8.2.1. *Quantification of Derivatized Uracil by GC-MS — Advantages and Concerns*

The first step in the analysis is release of genomic uracil by incubation with UNG. The polar nature of molecules like uracil requires it to be derivatised prior to GC-MS analysis to gain the minimum volatility necessary for GC separation. For uracil determination by GC-MS, derivatization has largely been carried out using 3,5-bis(trifluoromethyl) benzyl bromide (BTFMBzBr). A main advantage of this method is that UNG has high selectivity for uracil in DNA and that problems related to cytosine deamination during work-up is not a concern. The main disadvantage is limited yield of the derivatization reaction and instability of the derivatized molecules, e.g. due to sensitivity to moisture. GC-MS methods have been successfully used to quantify global uracil levels in genomic DNA[5,6,9–16] and it is less expensive and easier to operate than LC-MS/MS.

However, in addition to potential problems related to derivatization of uracil, there may also be concerns related to efficiency of uracil-release from DNA. The UNG-reaction is influenced by the varying ability of the enzyme to excise uracil in different genomic contexts. UNG efficiency is influenced by DNA conformation, sequence context, whether the DNA is single- or double-stranded, and whether the uracil is opposite a guanine or adenine.[17,18] Moreover, experimental conditions like Mg^{2+} and DNA concentration, which often vary between protocols, may affect excision efficiency (Chapter 4). Consequently, UNG-based assays may underestimate uracil levels by failing to excise all the substrate in the sample. A common control for UNG activity is the use of uracil-containing DNA, such as a PCR product amplified with dUTP instead of dTTP.[8] However, the uracil content in such PCR products is orders of magnitude higher than that of genomic DNA. Since the kinetics of UNG recruitment to uracil have not been compared in these contexts, it should not be assumed that UNG can detect and excise uracil in genomic DNA with the same efficiency as in short, uracil-dense DNA fragments. A strategy for ensuring complete uracil excision is to digest DNA with an endonuclease prior to UNG treatment,[6] which fragments the DNA and possibly increases enzyme substrate detection by minimizing supercoiling and lowering the DNA melting temperature, thereby increasing the accessibility to otherwise obscured uracil. Nevertheless, the aforementioned drawbacks of the UNG-based approach to uracil quantification are mostly theoretical in nature and largely amount to lacking confirmation that all uracils have been excised. Indeed, the convincing high quality data from several experiments that used this approach is a testament to its validity (discussed below).[6,19–21]

8.2.2. *Quantification of Deoxyuridine After Enzymatic Digestion of DNA to Deoxyribonucleosides*

Alternatively, genomic uracil can be quantified at the deoxynucleoside level after enzymatic hydrolysis of DNA by phoshodiesterases and phosphatases that cleave the phosphodiester backbone of DNA and dephosphorylate the resulting nucleotides, respectively. Although the risk of incomplete deoxyuridine release may be a concern it may be easier to control due to the existence of a number of different enzymes that can work efficiently together, many of which are relatively indiscriminate to the nature of the base modification, sequence context, DNA conformation, and experimental conditions. Indeed, enzymatic DNA hydrolysis and deoxyuridine quantification was found to be a more precise and reproducible assay for genomic uracil than the uracil excision approach when employed on a series of DNA samples isolated by various DNA extraction protocols.[4] The hydrolysis approach generally varied less than the uracil excision approach, especially when comparing different DNA isolation methods. The high variation in uracil levels measured using the uracil excision approach may indicate that the intrinsic UNG activity varies between samples. Alternatively, the higher variation could be explained by the fact that uracil levels were normalized to DNA measured spectrophotometrically, which is less precise than normalization to unmodified deoxyribonucleosides as in the hydrolysis approach. However, hydrolysing DNA to deoxyribonucleosides makes samples much more susceptible to *in vitro* cytosine deamination as described in the previous section, thereby introducing a possible over-estimation of uracil levels. When quantifying genomic uracil as liberated deoxyuridine by MS, co-purification with dUMP, dUDP or dUTP during DNA preparation also causes overestimation of genomic uracil because the nucleotides are dephosphorylated to deoxyuridine during DNA hydrolysis. Indeed, it was shown that nucleotide contamination in

DNA may occur when using common DNA purification techniques. However, up to 95% of total measured deoxyuridine in DNA preparations was removed when introducing a simple precipitation step, increasing to 98% if the DNA preparations were phosphatase-treated prior to precipitation. The latter observation could be explained by that the absence of a phosphate group slightly decreases co-precipitation of low molecular weight species with DNA.[4] MS-based analysis of deoxyuridine was also demonstrated to be complicated by the presence of 2′-deoxycytidine containing naturally occurring heavy isotopes of carbon, nitrogen, hydrogen and oxygen, of which [^{13}C] is the most frequently occurring. Approximately 9.9% of naturally-occurring 2′-deoxycytidine is 1 Da heavier than its monoisotopic molecular mass ([M+1]-deoxycytidine), and is therefore isobaric with and indistinguishable (by MS) from deoxyuridine. Although deoxyuridine and deoxycytidine are generally well separated using conventional chromatography, the deoxycytidine/deoxyuridine ratio in DNA is so high that the [M+1]-deoxycytidine peak tail still may obfuscate the deoxyuridine peak and subsequent MS analysis. A precursory, off-line chromatography step has been used as a means to prevent this problem by separating deoxyuridine from deoxycytidine prior to LC-MS/MS analysis.[2,4,22] The two chromatography steps have also been linked to make a two-dimensional LC-MS/MS method that can quantify a wide range of nucleosides, including deoxyuridine.[3]

Arguably, the most prominent error source during genomic uracil quantification by LC-MS/MS is the generation of uracil *in vitro* by cytosine deamination during sample preparation. Indeed, the fourth DNA nucleobase, thymine, is thought to be evolution's answer to the problem of the high rate of spontaneous cytosine deamination. Without thymine, cells could not differentiate between uracils incorporated by polymerases and those resulting from cytosine deamination[23-26] (Chapter 3). *In vitro* cytosine deamination during sample processing can be spontaneous or enzymatic. For practical purposes, cytosine deamination is only a problem when DNA is denatured to single-stranded DNA or nucleotides/nucleosides due to their vastly increased rates of cytosine deamination. Spontaneous cytosine deamination due to the intrinsic chemical instability of DNA is reportedly affected by three main factors: temperature, pH, and the degree to which DNA is denatured,[1,27,28] all of which typically vary between different analytical methods. For instance, some methods heat-denature DNA during sample preparation, using temperatures demonstrated to increase the uracil content up to 2.7-fold.[2,4,22,29] Furthermore, hydrolysis of DNA to single nucleosides at 37°C and pH 7.4 generated approximately 1.1 uracils per 10^8 cytosines per min, which is in accordance with previously reported values of cytosine deamination rates of 2.6, 0.12, and 0.0048 uracils per 10^8 cytosines per min for deoxyribonucleosides, single-stranded DNA, and double-stranded DNA, respectively.[4,28] The pH dependence of cytosine deamination was demonstrated as early as 1974 when Lindahl *et al.* measured *in vitro* deamination rates in DNA from *E. coli*,[1] The lowest deamination rate occurred at pH 7.5 to 8.5, with an especially strong increase above pH 9. Uracil may also arise through enzymatic cytosine deamination during sample workup. Cellular cytosine or cytidine deaminases may co-purify with genomic DNA or originate from other sources like recombinant enzyme preparations commonly used for sample treatment. The latter has been demonstrated to occur and thereby yield artificially high uracil and inosine values due to enzymatic cytosine and adenine deamination, respectively.[22] Taken together, the above factors demonstrate the importance of minimising high temperatures and pH extremes, and to include appropriate deaminase inhibitors during sample handling.

8.3. Other Methods to Quantify Genomic Uracil

8.3.1. *Radiometric Methods Detect Uracil in DNA*

Addition of radio-labelled uracil to cell cultures to quantify misincorporation into DNA or radio-labelled cytosine to measure deamination has been carried out as a part of wider experimental strategies. There are numerous examples of such approaches, e.g. using various treatments of cells, as well as examination of different mutants potentially affecting genomic uracil content. As examples, effects of 5-FU, as well as the role of SMUG1 and UNG were examined. Upon treatment with recombinant UNG and separation by HPLC, uracil was quantified by scintillation counting.[30,31] Such radiometric techniques may be useful, but their use is largely limited to cell cultures. Furthermore, metabolism of added bases or nucleosides may complicate interpretation (see Chapter 3).

8.3.2. *Laser-induced Fluorescence for Uracil Detection in DNA*

Laser-induced fluorescence (LIF) detection has been used to detect various endogenous lesions. LIF involves fluorophore-based derivatisation of the enzymatic digestion products of DNA prior to chromatographic separation and detection of the fluorescent label.[32] LIF detection is less sensitive, but more versatile than radiometric techniques since DNA from all sources may be analyzed. However, both radiometric and LIF detection can produce ambiguous results due to the lack of appropriate authentic standards and the fact that chromatographic retention is the only parameter that provides some chemical specificity to these techniques.

8.3.3. *Uracil-Binding Proteins to Determine Genomic Uracil*

Several chemical DNA modifications have been detected by using antibodies capable of recognising the DNA modification of interest.[33,34] A similar immunological approach is most likely not applicable to genomic uracil because sources of uracil as a base, nucleoside, and nucleotide are much more abundant than genomic uracil within any cell, requiring the DNA to be examined to have an unachievable degree of purity. Róna and colleagues instead created an indirect immunological detection system by generating a catalytically inactive UNG mutant with retained uracil-DNA binding capacity.[35] The UNG mutant, serving as a uracil sensor, was fused to epitope tags for immunological detection as well as to a fluorescent tag for direct fluorescent detection. The resulting dot-blot assay could detect uracil in various DNA with elevated uracil levels, including DNA from repair-deficient *E. coli* and a mismatch repair-deficient human cell line expressing the

UNG inhibitor, Ugi and treated with 5-fluorodeoxyuridine (5-FdU) to increase genomic uracil levels. However, the assay failed to detect UNG-dependent differences in genomic uracil in non-treated cells. In contrast, an alkoxyamine-based method (Section 8.1.2) detected 20–30-fold increased genomic uracil in Ung-deficient *E. coli* compared to wild-type, indicating either a sensitivity problem with the immune-detection method or that other glycosylases can maintain low uracil levels when UNG is inhibited in the 5-FdU-treated cell line.

8.4. Indirect Quantification of Uracil by Detecting UNG-induced Abasic Sites or Strand Breaks in DNA

Instead of detecting uracil itself, genomic uracil levels may be deduced by quantifying the abasic sites generated by UNG-mediated uracil excision or by quantifying single-strand breaks if uracil excision is carried out in combination with AP endonuclease 1 (APE1). One way to detect UDG-generated abasic sites in DNA is through specific reactions of the abasic sites with various alkoxyamines.

8.4.1. *Alkoxyamine Labelling of Abasic Sites and their Detection*

The alkoxyamine group reacts with the open-chain aldehyde form of abasic sites in DNA to produce open-chain aldehyde oximes (Fig. 8.3). Typically, the alkoxyamines used in abasic site assays also contain a moiety that may be used for conjugation to a detection probe, e.g. biotin,[36–38] or a methyl group containing a radioactive isotope.[39,40]

The biotin-tagged alkoxyamine, often referred to as Aldehyde-Reactive Probe (ARP), is available as part of a commercial kit (DNA Damage Quantification Kit, Dojindo Molecular Technologies). After tagging with ARP, basic-sites are quantified in an ELISA-like assay in which biotin is complexed with streptavidin, conjugated to the indicator enzyme horseradish peroxidase and detected colorimetrically. The ARP assay was developed further by combination with a slot blot technique in which the biotin-tagged DNA was spotted directly on a

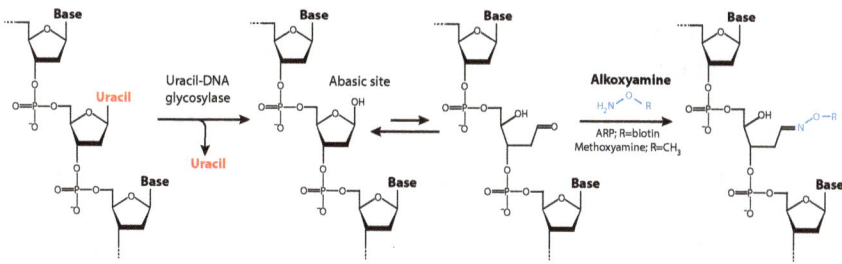

Fig. 8.3. Detection of uracil in DNA through labelling abasic sites with alkoxy-amines. Following *in vitro* excision of uracil by UNG, the open-chain aldehyde form of the resulting abasic sites are reacted with alkoxyamines to produce aldehyde oximes, thereby introducing a label that may be used for detection.

membrane prior to detection.[41] Although the slot blot version of the assay involved the same colorimetric detection through streptavidin-conjugated horseradish peroxidase, the slot blotting improved the assay sensitivity 20-fold from 5 to 0.24 abasic sites per 10^6 bases compared to the commercial kit, most of which was due to increased sample sizes (from 1 to 15 µg DNA per sample).[41] Yet another version was developed by Shalhout *et al.*,[21] who rather than colorimetric detection employed conjugation with streptavidin-Cy5 and fluorescence detection of the ARP-tagged, membrane-bound DNA.[21] While several research groups have found ARP assays useful for the quantification of abasic sites in DNA,[21,36,38,42] the method is hampered by several limitations. First, alkoxyamines react not only with abasic sites, but also with all ketone- and aldehyde-containing moieties in DNA. For instance, deoxyribose oxidation produces a variety of ketone- and aldehyde-containing products that potentially contribute to the signal recorded by ARP assays.[40,43] Second, noncovalent intercalation of ARP with DNA may occur, resulting in additional overestimation of abasic sites. Third, the colorimetric assay used to quantify ARP-DNA has an only 40-fold dynamic range, which is rather limited for use in a biological context. Finally, the reaction of DNA with ARP and other alkoxyamines is performed most effectively at acidic pH at which extensive spontaneous depurination occurs, which could generate significant amounts of abasic sites in addition to the ones originally present in DNA.[44,45] If the DNA — ARP reaction is buffered to neutral pH, it would likely be incomplete, leaving abasic sites unreacted and leading to underestimation of abasic sites. Indeed, ARP was found to react with less than 20% of the abasic sites in an oligomer at pH 7 or 8.[45] The quantitative accuracy of ARP based assays has therefore been questioned.[40,45] Several laboratories have adapted alkoxyamine-based abasic site assays to quantify genomic uracil, through pre-treatment of DNA with UNG.[21,38,46] Cabelof and colleagues (2006) were the first to employ this assay extension in combination with ARP to monitor changes in genomic uracil in mice fed varying amounts of dietary folate.[46] Well-established characteristics of folate-depleted cells include increased dUTP to dTTP ratio, which in turn leads to increased uracil misincorporation into DNA (Chapter 3.1).[47,48] In line with this, the UNG-ARP assay detected a small but significant increase in genomic uracil in folate deficient mice.[46] Using a similar approach, the steady-state levels of uracil in DNA of *E. coli* strains were measured in order to determine the influence of Ung, G/U mismatch uracil-DNA glycosylase (Mug), and dUTP pyrophosphatase (Dut) on uracil accumulation. Although genomic uracil levels in the wild-type strain were at the limit of detection of approximately 1 uracil per million bases, 20- to 30-fold increases were detected in Ung- and Ung/Dut-deficient strains, demonstrating the usefulness of the assay.[38] Furthermore, Shalhout *et al.* were able to detect an 11-fold increase in genomic uracil levels in UNG-deficient murine B-lymphocytes following stimulation,[21] which is qualitatively consistent with the increased uracil levels in the immunoglobulin genes of stimulated UNG-deficient B-cells as detected by a PCR-based method (Section 8.1.3).[49] Furthermore, they found a correlation between activation-induced cytidine deaminase (AID) expression and uracil levels in different B-cell lymphoma cell lines that mirrored the results from an MS-based assay that quantified deoxyuridine.[21,50]

8.4.2. *The Comet Assay to Detect Genomic Uracil*

The comet assay (also termed the single-cell gel electrophoresis assay) is a simple, rapid, sensitive, and extensively used method for the relative quantification of DNA strand breaks and other DNA lesions in individual eukaryotic cells. It involves the encapsulation of cells in an agarose suspension, cell lysis, DNA unwinding in neutral or alkaline conditions, and electrophoresis allowing possible broken DNA fragments to migrate away from the nucleus. The migrated DNA often resembles a comet tail, hence the term comet assay, and the portion of DNA within the comet tail is directly proportional to the amount of DNA damage. Although the damages detected are DNA strand breaks, adjustments to the protocol enables the comet assay to be used to quantify the presence of a wide variety of DNA lesions, including abasic sites, oxidized bases, and uracil. This is possible with the inclusion of enzymatic digestion with specific DNA glycosylase/endonuclease enzymes that convert the lesions into single-strand breaks. To detect uracil, a UDG and APE1 are employed to excise uracil and cleave the resulting abasic site, respectively.[51,52] The uracil-sensitive comet assay was successfully used to show increased uracil misincorporation in folate-deprived lymphocytes,[52,53] as well as elevated genomic uracil levels in MEF cells derived from $Ung^{-/-}$ mice and lymphoid cell lines derived from UNG mutant patients.[54,55] However, the comet assay has its limitations. The damage quantification is usually only relative and the variation is typically high, the latter making it necessary to analyze a large number of cells to achieve statistical validity. Moreover, although the assay is very sensitive (detection limit estimated to about 1 break per 15 million base pairs), only a relatively narrow range of break frequencies can be detected because very small DNA fragments will diffuse and become lost even before electrophoresis starts. Thus, closely spaced lesions would be severely underrepresented. Nevertheless, the assay has proven to be a very valuable tool in the study of DNA damage and repair, particularly in that it makes it possible to assess intercellular differences. Moreover, extremely small cell samples are required (one to ten thousand cells), and the assay can be adapted to many different kinds of DNA damage.

A major drawback of any method for uracil detection through UNG-generated abasic sites or strand breaks is that they all determine uracil levels by subtracting the baseline levels of abasic sites or strand breaks in DNA. During conditions in which abasic sites or single-strand breaks occur at high levels (e.g. in response to DNA-damaging agents), the background noise may thus be too high for accurate detection of uracil. Thus, these assays should only be considered semi-quantitative during such conditions.

8.5. Methods to Determine Where in the Genome Uracil is Localized

The most informative uracil measurement would involve both accurate quantification and localization, but current assays that give localization data are semi-quantitative at best. To date four general approaches have been utilized to localize genomic uracil (Fig. 8.4).

8.5.1. *Quantification and Localisation of Genomic Uracil by PCR-based Methods*

The first approach was employed by Maul and colleagues and involves UNG- and APE1-treatment of DNA prior to either Southern-blot analysis, and/or ligation-mediated PCR.[49] Here, UNG excises uracils from the genome and APE1 cleaves the resulting abasic sites. UNG/APE1-treated DNA could then be analyzed by Southern-blot using probes complementary to the genomic regions of interest. The chicken DT40 cell line was used, which undergoes constitutive AID-dependent Ig gene conversion, and it is widely used to study antibody diversification as well as DNA repair. Since one allele is inactive in DT40 cells, the non-rearranged allele could be used as a loading control in the Southern blots. The same group also employed a second approach, in which UNG/APE1-treated DNA was subjected to quantitative PCR (qPCR). The ratios of amplified product from treated vs. untreated DNA was then used to estimate uracil content in immunoglobulin genes.[49]

Another group employed a similar technique, but used end-product qPCR instead of real-time PCR, which likely decreased the sensitivity of their assay.[48] Alternatively, UNG/APE1-treated DNA was also processed with a mutant form of DNA polymerase β that lacked polymerase activity but retained lyase activity, to process the nick left

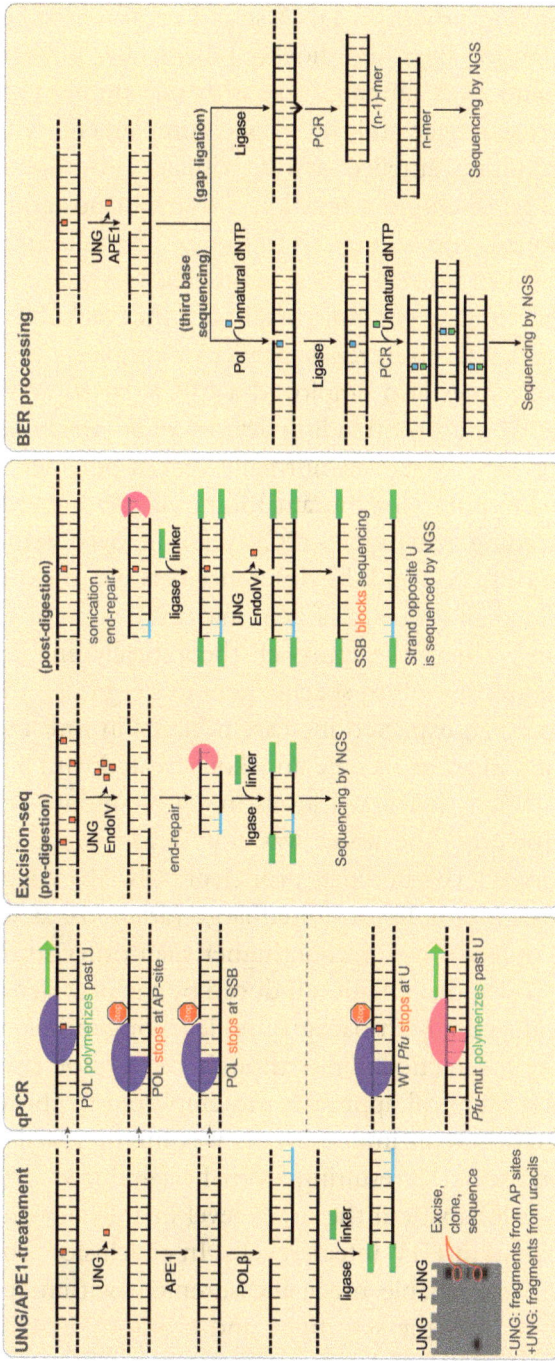

Fig. 8.4. Overview of the methods used to determine uracil levels within specific gene regions. Uracil can be excised from DNA by treatment with a uracil DNA glycosylase and the resulting AP site cleaved with an AP endonuclease. Downstream, the nicked DNA can be compared to untreated DNA by qPCR or processed with polymerases[49,56] and/or ligase to yield sequenceable fragments.[49,57-59] Alternatively, qPCR can be performed using a *Pfu* mutant lacking uracil proofreading capacity and compared to wild-type *Pfu*.[60]

by APE1.[49] The resulting processed DNA strand break was subjected to one round of extension by a high-fidelity polymerase, followed by ligation to a linker and PCR amplification with one primer complementary to the linker sequence and the second complementary to the target gene. The resulting amplicons were visualised by Southern-blotting (each band representing a uracil at a different position) and could be excised, cloned, and sequenced to localize the uracil responsible for each band. Taken together, these approaches provide a means to semi-quantitatively localize genomic uracil, but they lack the sensitivity to detect low uracil levels in repair-proficient samples.

Horvath and colleagues also employed a qPCR-based approach by taking advantage of the proofreading activity of *P. furiosus* (*Pfu*) on uracil-containing DNA.[60] Uracil normally blocks polymerisation by *Pfu* polymerase (Pfu pol), so a mutant Pfu pol was employed that lacked uracil proofreading activity.[61,62] qPCR was employed with both wild-type and mutant Pfu pol, and the difference between the outputs of each reaction yielded an estimate of genomic uracil content within the amplified regions. Thus, this approach theoretically enables the study of the relative uracil levels in specific genomic regions. Analysis of *E. coli* DNA showed an expected increase in uracil in *ung* and *dut* double knockouts and in an *ung* single knockout treated with 5-FdU. The same treatment increased uracil levels in *Ung*[-/-] mouse embryonic fibroblasts. However, the assay was not sensitive enough to detect variation in uracil levels in repair-proficient cells. Moreover, the results from mammalian samples exhibited large variations, as well as uracil levels several orders of magnitude higher than other studies.[4,7] It is possible that cytosine deamination induced by the heat denaturation PCR step introduced a high background level of genomic uracil that raised the assay's detection limit, though this is not always a problem as an almost identical approach was employed to show that HIV DNA in infected human immune cells was highly uracilated.[63] That is, the high noise level resulting from heat-induced cytosine deamination is less problematic if the uracil level in a given sample is sufficiently high, such as in HIV-infected cells. In conclusion, the Pfu pol qPCR approach may enable uracil measurement within specific genomic regions, but lacks the sensitivity and precision required to measure subtle changes in uracil levels.

Similarly, a well-established telomere length qPCR assay has also been slightly modified to estimate the uracil content of telomeric DNA combined with the above qPCR uracil assay.[56,64] Telomeric DNA was initially difficult to amplify by PCR because the TTAGGG repeats present in telomeres required complementary primers that would form dimmers.[65] The original telomere length assay circumvents this problem by relying on primers that bind to C- and G-rich DNA segments while mismatching at the other bases as well as by using low annealing temperature during the first two PCR cycles.[66] The low annealing temperature allows binding of the primers to telomeric DNA despite the mismatches. After two PCR cycles, the annealing temperature is increased to allow annealing of the primers only to the recently-generated PCR amplicons that share full sequence complementarity with the primers. Absolute quantification of telomere length is achieved by normalizing the qPCR results to parallel qPCRs with an oligo of fixed size and that contains the telomeric repeat sequence and a single-copy gene (*e.g. GAPDH*).[67,68] Finally, to measure uracil levels, samples are treated with UNG or UNG and APE1 to excise uracils and nick the resulting AP site prior to qPCR.[56,64] The resulting abasic sites or nicks in the template strand block the DNA polymerase, so UNG/APE1-treated samples can be compared to untreated samples to estimate the amount of uracil present in telomeres. qPCR of UNG/APE1-treated DNA was used to show uracil accumulation in telomeric DNA in UNG-deficient mouse embryonic fibroblasts.[56] Moreover, qPCR of UNG-treated DNA was used to show that culturing cells in folate-deficient medium to increase the dUTP/dTTP pool also led to telomeric uracil accumulation.[64] Since very few samples were studied, it is difficult to comment on the precision and reproducibility of this approach, but as a PCR-based assay, it shares similar drawbacks to the aforementioned methods.

8.5.2. *Uracil Localization by Ligating Adapters to Cleaved Abasic Sites and Sequencing*

In the different approach for quantifying local uracil levels, Bryan and colleagues developed a method to localize uracil with single-base resolution.[57] The strategy involved treating DNA with UNG to excise uracils and then endonuclease IV to nick the resulting abasic sites, leaving single base gaps that were end-repaired and ligated to adaptors for next-generation high-resolution sequencing. The first base after the adaptors would correspond to uracils, thereby enabling their localization. The major disadvantage with this strategy is that it requires a high uracil content to yield ligatable fragments for sequencing library construction. Thus, another approach was developed, in which DNA was sheared and then ligated to adapters prior to UNG-treatment. After the adapters were ligated, samples were treated with UNG and endonuclease IV to nick the uracil-containing strand, thereby rendering it incompatible for sequencing and enabling a less precise uracil localization by sequencing the unmodified strand (and potentially also by comparing the results to those obtained by the pre-ligation UNG digestion approach). In this case, the assay cannot detect uracils when they are proximal but on different strands.

8.5.3. *Exploiting the BER Pathway to Localize Genomic Uracil*

The fourth approach to measure local uracil content relies on appropriating the BER pathway.[58,59] Riedl and colleagues presented two methods in which DNA was initially treated with UNG and APE1 to generate DNA nicks. The two methods then diverged. In the first method, an unnatural nucleotide (dNaM or d5SICS) was inserted by DNA polymerase.[58] Next, the nick was sealed with a ligase and the sealed duplex was PCR amplified with another unnatural nucleotide (d5SICS or dMMO2) that paired with the originally inserted unnatural nucleotide. The amplified PCR product could later be sequenced with single-molecule sequencing (so-called Nanopore sequencing). In this method, the ligation step was found to have low reaction yield (~56–72%). Thus, an alternative polymerisation step was proposed, in which a strand-displacing polymerase was used to insert the unnatural base and extend the strand. However, this strategy only allowed for one lesion to be detected per strand. Furthermore, PCR amplification of lesion-free DNA containing only the four canonical bases is more efficient than that PCR of DNA containing unnatural bases, so the latter must be purified to enrich for PCR product containing the unnatural bases, limiting the usefulness of the assay in a biological context. Lastly, unnatural bases cannot yet be sequenced by available deep sequencing technologies, although Oxford Nanopore's MinION platform may soon accomplish this.[69,70]

The second method proposed by this group omitted the use of unnatural bases by directly ligating single-strand breaks produced by UNG and APE1.[59] The resulting DNA duplex can be PCR-amplified and sequenced with more conventional technologies, yielding a product strand missing a base at the position at which the lesion was removed. Here again, the ligase reaction exhibited a low yield of ~45–50%, depending on whether ligating single or multiple lesions. Moreover, many consecutive uracils would yield many deletions, so detection of the exact locations or the uracils would be limited in this context. Finally, deletions are common artefacts in common deep sequencing methods,[70,71] so this method requires either an extraordinarily high sequencing depth for whole genome coverage or should be used in targeted regions. A similar approach employs co-amplification at lower denaturation temperature PCR (COLD-PCR) post-ligation to selectively amplify the bulge in the uracil-containing DNA, although their claimed sensitivity of 0.01% is unlikely to detect basal uracil levels.[72]

In summary, a problem common to all uracil sequencing approaches is that of resolution. That is, whichever sequencing method is used it must produce enough reads to detect the lesion in one out of more than 10^7 bases. The need for such a high resolution arises because even though uracil levels have been reported to be around one per 10^6 bases, the likelihood that the uracils are found in the same position is low. Furthermore, any meaningful study must also differentiate between induced uracils (from treatment or genotype, for instance) and background uracil levels. Thus, very high resolution (and sequencing fidelity) is needed to detect uracils above the background. Current sequencing technologies therefore either require an enrichment step prior to uracil sequencing or a high abundance of the lesion.

8.6. Reported Values for Genomic Uracil and Conclusions

The reported values for basal genomic uracil levels in uracil repair-competent mammalian cells are highly divergent, ranging from 0.02 to 345 uracil moieties per million bases (Table 8.1). Although this large variation may partially depend on the cell type and conditions studied, it may also be largely related to the technical problems associated with reliable quantification of genomic uracil. Current methods often detect genomic uracil at or near the detection limit, where the assay is least accurate. All methods also rely on some sort of normalization. However, even when using standard curves as well as stable isotope-labelled standards, true absolute quantification and inter-experimental comparability are difficult to achieve. Furthermore, various technical issues detailed above plague all approaches, including the underestimation of uracil content from incomplete UNG excision and its overestimation from *in vitro* cytosine deamination.

The most appropriate approach to quantify genomic uracil depends on the experiment, as well as the available equipment. For global uracil quantification, measuring uracil or deoxyuridine by mass spectrometry will yield the most specific, accurate and precise data, but can be time-consuming to set up and requires expensive instrumentation. Approaches that hydrolyze DNA to nucleosides to measure deoxyuridine and treat DNA with UNG to analyze uracil have been employed to successfully measure basal uracil levels, and both approaches have significant advantages and drawbacks. A major advantage to the DNA hydrolysis strategy is the ability to concurrently measure other nucleoside modifications, as well as a more accurate normalization to unmodified nucleosides. Excluding the use of mass spectrometry, using UNG to excise uracil followed by labelling of the resulting abasic site with a biotinylated probe has shown basal genomic uracil levels that correlate well with independent experiments employing LC-MS/MS.[21,50] Using a catalytically inactive form of UNG as a uracil detector is a simpler and more elegant approach, but use of this method showed no difference in uracil levels after treatment with 5-FdU, and may thus lack the sensitivity required for measuring subtle differences in uracil levels.[35] For local genomic uracil determination, the abundance and proximity of

Table 8.1. Analytical Methods that Detect Uracil in DNA and Reported Basal Levels in Repair-proficient Cells

Method	Sample Type	U/10⁶ dN	Ref.
Global levels, direct measurement			
2D LC-MS/MS	human B-lymphoblast cell line	10.5	22
2D LC-MS/MS	human B-lymphoblast cell line, MEFs	0.072–0.105	4
2D LC-MS/MS	human B-cells and B-cell lines, mouse B-cell line	0.056–4.03	50
2D LC-MS/MS	human colon and colon adenocarcinoma	9.3–14.6	3
2D LC-MS/MS	mouse tissues	1.14–4.73	73
LC-MS	human liver carcinoma	8.78	29
LC-MS/MS	human lymphocytes	0.55–5.34	8
LC-MS/MS	human embryonic kidney cell line	12	74
LC-MS/MS	colon cancer cell line	A.U.	19
LC-MS/MS	human breast cancer cell lines	20	20
LC-UV	human bone marrow	113.9–118.1	75
CE-LIF	human leukaemia cells	50–80	32
GC/MS	rat hepatocytes	1.9–8.3	9
GC/MS	human peripheral leukocytes	4.57–354.8	10
GC/MS	mouse erythroblasts	1.27–4.87	15
GC/MS	human lymphocytes	2.09–13.3	12
GC/MS	rat colon	10.7–14	11
GC/MS	human lymphocytes	0.91–17.5	5
GC/MS	human lymphocytes	0.27–0.4	6
GC/MS	mouse liver	0.36–1.6	16
GC/MS	human fibroblasts	1.35–1.57	13
GC/MS	human cell lines, MEFs	0.14–8.3	14
UDG-X, fluorescence	*E. coli*	N.A.	76
UNG-mut, fluorescence	colon cancer cell line	10.5	35
Global levels, indirect measurement			
alkaline elution, comet	mouse hepatocytes, splenocytes, spermatozoa	0.02–0.05	77

(*Continued*)

Table 8.1. (*Continued*)

Method	Sample Type	U/10⁶ dN	Ref.
comet	human cell lines	A.U.	52
comet	human lymphocytes	A.U.	53
comet	human B-cells	A.U.	54
ARP, fluorescence	human, mouse B-cells and cell lines	A.U.	21
ARP, fluorescence	mouse liver, spleen, colon	230–345	46
ARP, fluorescence	*E. coli*	N.A.	38
Local levels			
Pfu-mut qPCR	MEFs	N.D.	60
Pfu-mut qPCR	HIV in infected human leukocytes	A.U.	63
UNG-treat, gap ligation PCR	plasmids	A.U.	59
UNG-treat, qPCR	rat liver	A.U.	48
UNG-treat, qPCR, LM-PCR, Southern	chicken B-cell line, mouse splenocytes	A.U.	49
UNG-treat, unnatural base insertion	plasmids	A.U.	58
UNG-treat, Excision-seq	budding yeast	A.U.	57
UNG-treat, telomere qPCR	human lymphoblasts	A.U.	64

A.U. = arbitrary units, N.A. = not applicable (e.g. basal levels were not reported).

the uracils are the main deciding factors when selecting a method. Measuring differences using qPCR (both with UNG-treated DNA and using mutant Pfu pol) requires large differences between samples. Furthermore, uracil measurement by qPCR is best employed when a localized uracilation event is expected, such as deamination by AID at the immunoglobulin gene loci, since a non-uracilated gene can be used to normalize between samples. True genome-wide uracil sequencing is not yet achievable because no method can fully localize all uracils during sequencing. Moreover, even if a method could sequence genomic uracil, the resolution of the method would have to be very much higher than currently available or the uracils would need to be in the same location for all cells. Thus, an enrichment step prior to sequencing may be required to sequence uracils genome-wide. The current best option for single-base resolution uracil localization is LM-PCR followed by cloning and sequencing of the resulting amplicons, though this approach only yielded usable results in UNG-deficient cells and also requires a certain distance between uracils to yield clonable amplicons.[49]

In conclusion, accurately quantifying uracil is challenging and no single approach can overcome all the technical complications. Furthermore, a method for localizing genomic uracil remains elusive, as current methods only provide rough positional data and are prone to more technical complications than global uracil assays. Nevertheless, an assay that provides accurate genomic uracil quantification as well as single-base-resolution positioning is crucial to understanding uracil's role in DNA *in situ*. Modern molecular biological techniques, especially sequencing technologies, are in a constant state of rapid advancement, and there is a significant amount of ongoing work relating to genomic uracil, so such an assay is likely to be developed in the near future.

References

1. Lindahl, T., Nyberg, B. Heat-induced deamination of cytosine residues in deoxyribonucleic acid. *Biochemistry* **13**, 3405–3410 (1974).
2. Dong, M., Wang, C., Deen, W. M., Dedon, P. C. Absence of 2′-deoxyoxanosine and presence of abasic sites in DNA exposed to nitric oxide at controlled physiological concentrations. *Chem Res Toxicol* **16**, 1044–1055 (2003).
3. Gackowski, D., Starczak, M., Zarakowska, E., *et al.* Accurate, Direct, and High-Throughput Analyses of a Broad Spectrum of Endogenously Generated DNA Base Modifications with Isotope-Dilution Two-Dimensional Ultraperformance Liquid Chromatography with Tandem Mass Spectrometry: Possible Clinical Implication. *Anal Chem* **88**, 12128–12136 (2016).

4. Galashevskaya, A., Sarno, A., Vagbo, C. B., Aas *et al*. A robust, sensitive assay for genomic uracil determination by LC/MS/MS reveals lower levels than previously reported. *DNA Repair (Amst)* **12**, 699–706 (2013).

5. Mashiyama, S. T., Courtemanche, C., Elson-Schwab, I., *et al*. Uracil in DNA, determined by an improved assay, is increased when deoxynucleosides are added to folate-deficient cultured human lymphocytes. *Anal Biochem* **330**, 58–69 (2004).

6. Mashiyama, S. T., Hansen, C. M., Roitman, E., *et al*. An assay for uracil in human DNA at baseline: effect of marginal vitamin B6 deficiency. *Anal Biochem* **372**, 21–31 (2008).

7. Olinski, R., Jurgowiak, M., Zaremba, T. Uracil in DNA-Its biological significance. *Mutat Res* **705**, 239–245 (2010).

8. Ren, J., Ulvik, A., Refsum, H., Ueland, P. M. Uracil in human DNA from subjects with normal and impaired folate status as determined by high-performance liquid chromatography-tandem mass spectrometry. *Anal Chem* **74**, 295–299 (2002).

9. Blount, B. C., Ames, B. N. Analysis of uracil in DNA by gas chromatography-mass spectrometry. *Anal Biochem* **219**, 195–200 (1994).

10. Blount, B. C., Mack, M. M., Wehr, C. M., *et al*. Folate deficiency causes uracil misincorporation into human DNA and chromosome breakage: implications for cancer and neuronal damage. *Proc Natl Acad Sci U S A* **94**, 3290–3295 (1997).

11. Choi, S. W., Friso, S., Dolnikowski, G. G., *et al*. Biochemical and molecular aberrations in the rat colon due to folate depletion are age-specific. *J Nutr* **133**, 1206–1212 (2003).

12. Crott, J. W., Mashiyama, S. T., Ames, B. N., Fenech, M. F. Methylenetetrahydrofolate reductase C677T polymorphism does not alter folic acid deficiency-induced uracil incorporation into primary human lymphocyte DNA *in vitro*. *Carcinogenesis* **22**, 1019–1025 (2001).

13. Field, M. S., Kamynina, E., Watkins, D., *et al*. Human mutations in methylene-tetrahydrofolate dehydrogenase 1 impair nuclear de novo thymidylate biosynthesis. *Proceedings of the National Academy of Sciences* **112**, 400–405 (2015).

14. Kamynina, E., Lachenauer, E. R., DiRisio, A. C., *et al*. Arsenic trioxide targets MTHFD1 and SUMO-dependent nuclear de novo thymidylate biosynthesis. *Proc Natl Acad Sci U S A* **114**, E2319–E2326 (2017).

15. Koury, M. J., Horne, D. W., Brown, Z. A., *et al*. Apoptosis of late-stage erythroblasts in megaloblastic anemia: association with DNA damage and macrocyte production. *Blood* **89**, 4617–4623 (1997).

16. MacFarlane, A. J., Liu, X., Perry, C. A., *et al*. Cytoplasmic Serine Hydroxymethyltransferase Regulates the Metabolic Partitioning of Methylenetetrahydrofolate but Is Not Essential in Mice. *J Biol Chem* **283**, 25846–25853 (2008).

17. Kavli, B., Sundheim, O., Akbari, M., *et al.* hUNG2 is the major repair enzyme for removal of uracil from U:A matches, U:G mismatches, and U in single-stranded DNA, with hSMUG1 as a broad specificity backup. *J Biol Chem* **277**, 39926–39936 (2002).

18. Slupphaug, G., Eftedal, I., Kavli, B., *et al.* Properties of a recombinant human uracil-DNA glycosylase from the UNG gene and evidence that UNG encodes the major uracil-DNA glycosylase. *Biochemistry* **34**, 128–138 (1995).

19. Bulgar, A. D., Weeks, L. D., Miao, Y., *et al.* Removal of uracil by uracil DNA glycosylase limits pemetrexed cytotoxicity: overriding the limit with methoxy-amine to inhibit base excision repair. *Cell Death Dis* **3**, e252 (2012).

20. Burns, M. B., Lackey, L., Carpenter, M. A., *et al.* APOBEC3B is an enzymatic source of mutation in breast cancer. *Nature* **494**, 366–370 (2013).

21. Shalhout, S., Haddad, D., Sosin, A., *et al.* Genomic uracil homeostasis during normal B cell maturation and loss of this balance during B cell cancer development. *Mol Cell Biol* **34**, 4019–4032 (2014).

22. Dong, M., Dedon, P. C. Relatively small increases in the steady-state levels of nucleobase deamination products in DNA from human TK6 cells exposed to toxic levels of nitric oxide. *Chem Res Toxicol* **19**, 50–57 (2006).

23. Dube, D. K., Kunkel, T. A., Seal, G., Loeb, L. A. Distinctive properties of mammalian DNA polymerases. *Biochim Biophys Acta* **561**, 369–382 (1979).

24. Focher, F., Mazzarello, P., Verri, A., *et al.* Activity profiles of enzymes that control the uracil incorporation into DNA during neuronal development. *Mutat Res* **237**, 65–73 (1990).

25. Kunkel, T. A., Loeb, L. A. On the fidelity of DNA replication. Effect of divalent metal ion activators and deoxyrionucleoside triphosphate pools on *in vitro* mutagenesis. *J Biol Chem* **254**, 5718–5725 (1979).

26. Mazzarello, P., Focher, F., Verri, A., Spadari, S. Misincorporation of uracil into DNA as possible contributor to neuronal aging and abiotrophy. *The International Journal of Neuroscience* **50**, 169–174 (1990).

27. Lindahl, T. Instability and decay of the primary structure of DNA. *Nature* **362**, 709–715 (1993).

28. Shapiro, R. in *Chromosome Damage and Repair NATO Advanced Study Institutes Series* (ed. Erling Seeberg, Kjell Kleppe) 3–18 (Springer US, 1981).

29. Chango, A., Abdel Nour, A. M., Niquet, C., Tessier, F. J. Simultaneous determination of genomic DNA methylation and uracil misincorporation. *Med Princ Pract* **18**, 81–84 (2009).

30. An, Q., Robins, P., Lindahl, T., Barnes, D. E. C --> T mutagenesis and gamma-radiation sensitivity due to deficiency in the Smug1 and Ung DNA glycosylases. *EMBO J* **24**, 2205–2213 (2005).

31. An, Q., Robins, P., Lindahl, T., Barnes, D. E. 5-Fluorouracil incorporated into DNA is excised by the Smug1 DNA glycosylase to reduce drug cytotoxicity. *Cancer Res* **67**, 940–945 (2007).

32. Wirtz, M., Schumann, C. A., Schellenträger, M., *et al.* Capillary electrophoresis-laser induced fluorescence analysis of endogenous damage in mitochondrial and genomic DNA. *Electrophoresis* **26**, 2599–2607 (2005).

33. Chowdhury, B., Cho, I.-H., Hahn, N., Irudayaraj, J. Quantification of 5-methylcytosine, 5-hydroxymethylcytosine and 5-carboxylcytosine from the blood of cancer patients by an enzyme-based immunoassay. *Anal Chim Acta* **852**, 212–217 (2014).

34. Leadon, S. A. Production of thymine glycols in DNA by radiation and chemical carcinogens as detected by a monoclonal antibody. *The British Journal of Cancer Supplement* **8**, 113–117 (1987).

35. Róna, G., Scheer, I., Nagy, K., *et al.* Detection of uracil within DNA using a sensitive labeling method for *in vitro* and cellular applications. *Nucleic Acids Res* **44**, e28 (2016).

36. Atamna, H., Cheung, I., Ames, B. N. A method for detecting abasic sites in living cells: age-dependent changes in base excision repair. *Proc Natl Acad Sci USA* **97**, 686–691 (2000).

37. Kubo, K., Ide, H., Wallace, S. S., Kow, Y. W. A novel, sensitive, and specific assay for abasic sites, the most commonly produced DNA lesion. *Biochemistry* **31**, 3703–3708 (1992).

38. Lari, S. U., Chen, C. Y., Vertessy, B. G., *et al.* Quantitative determination of uracil residues in Escherichia coli DNA: Contribution of ung, dug, and dut genes to uracil avoidance. *DNA Repair (Amst)* **5**, 1407–1420 (2006).

39. Wang, Y., Liu, L., Wu, C., *et al.* Direct detection and quantification of abasic sites for *in vivo* studies of DNA damage and repair. *Nucl Med Biol* **36**, 975–983 (2009).

40. Zhou, X., Liberman, R. G., Skipper, P. L., *et al.* Quantification of DNA strand breaks and abasic sites by oxime derivatization and accelerator mass spectrometry: application to gamma-radiation and peroxynitrite. *Anal Biochem* **343**, 84–92 (2005).

41. Nakamura, J., Walker, V. E., Upton, P. B., *et al.* Highly sensitive apurinic/apyrimidinic site assay can detect spontaneous and chemically induced depurination under physiological conditions. *Cancer Res* **58**, 222–225 (1998).

42. Nakamura, J., Swenberg, J. A. Endogenous apurinic/apyrimidinic sites in genomic DNA of mammalian tissues. *Cancer Res* **59**, 2522–2526 (1999).

43. Pogozelski, W. K., Tullius, T. D. Oxidative strand scission of nucleic acids: routes initiated by hydrogen abstraction from the sugar moiety. *Chem Rev* **98**, 1089–1108 (1998).

44. Shapiro, R. in *Chromosome Damage and Repair, Ed's Erling Seeberg and Kjell Kleppe* Vol. 40, 3–18 (Plenum Press, ISBN 0-306-40886-4, New York London, 1980).

45. Wei, S., Shalhout, S., Ahn, Y.-H., Bhagwat, A. S. A versatile new tool to quantify abasic sites in DNA and inhibit base excision repair. *DNA Repair 27*, 9–18 (2015).

46. Cabelof, D. C., Nakamura, J., Heydari, A. R. A sensitive biochemical assay for the detection of uracil. *Environ Mol Mutag* **47**, 31–37 (2006).

47. James, S. J., Miller, B. J., Basnakian, A. G., *et al.* Apoptosis and proliferation under conditions of deoxynucleotide pool imbalance in liver of folate/methyl deficient rats. *Carcinogenesis* **18**, 287–293 (1997).

48. Pogribny, I. P., Muskhelishvili, L., Miller, B. J., James, S. J. Presence and consequence of uracil in preneoplastic DNA from folate/methyl-deficient rats. *Carcinogenesis* **18**, 2071–2076 (1997).

49. Maul, R. W., Saribasak, H., Martomo, S. A., *et al.* Uracil residues dependent on the deaminase AID in immunoglobulin gene variable and switch regions. *Nat Immunol* **12**, 70–76 (2011).

50. Pettersen, H. S., Galashevskaya, A., Doseth, B., *et al.* AID expression in B-cell lymphomas causes accumulation of genomic uracil and a distinct AID mutational signature. *DNA Repair (Amst)* **25**, 60–71 (2015).

51. Collins, A. R., Duthie, S. J., Dobson, V. L. Direct enzymic detection of endogenous oxidative base damage in human lymphocyte DNA. *Carcinogenesis* **14**, 1733–1735 (1993).

52. Duthie, S. J., McMillan, P. Uracil misincorporation in human DNA detected using single cell gel electrophoresis. *Carcinogenesis* **18**, 1709–1714 (1997).

53. Duthie, S. J., Hawdon, A. DNA instability (strand breakage, uracil misincorporation, and defective repair) is increased by folic acid depletion in human lymphocytes *in vitro*. *FASEB journal: official publication of the Federation of American Societies for Experimental Biology* **12**, 1491–1497 (1998).

54. Kavli, B., Andersen, S., Otterlei, M., *et al.* B cells from hyper-IgM patients carrying UNG mutations lack ability to remove uracil from ssDNA and have elevated genomic uracil. *J Exp Med* **201**, 2011–2021 (2005).

55. Nilsen, H., Rosewell, I., Robins, P., *et al.* Uracil-DNA glycosylase (UNG)-deficient mice reveal a primary role of the enzyme during DNA replication. *Mol Cell* **5**, 1059–1065 (2000).

56. Vallabhaneni, H., Zhou, F., Maul, R. W., *et al.* Defective repair of uracil causes telomere defects in mouse hematopoietic cells. *J Biol Chem* **290**, 5502–5511 (2015).

57. Bryan, D. S., Ransom, M., Adane, B., *et al.* High resolution mapping of modified DNA nucleobases using excision repair enzymes. *Genome Res* **24**, 1534–1542 (2014).

58. Riedl, J., Ding, Y., Fleming, A. M., Burrows, C. J. Identification of DNA lesions using a third base pair for amplification and nanopore sequencing. *Nature Communications* **6**, 8807 (2015).

59. Riedl, J., Fleming, A. M., Burrows, C. J. Sequencing of DNA Lesions Facilitated by Site-Specific Excision via Base Excision Repair DNA Glycosylases Yielding Ligatable Gaps. *J Am Chem Soc* **138**, 491–494 (2016).

60. Horváth, A., Vértessy, B. G. A one-step method for quantitative determination of uracil in DNA by real-time PCR. *Nucleic Acids Res* **38**, e196 (2010).

61. Fogg, M. J., Pearl, L. H., Connolly, B. A. Structural basis for uracil recognition by archaeal family B DNA polymerases. *Nat Struct Biol* **9**, 922–927 (2002).

62. Gill, S., O'Neill, R., Lewis, R. J., Connolly, B. A. Interaction of the family-B DNA polymerase from the archaeon Pyrococcus furiosus with deaminated bases. *J Mol Biol* **372**, 855–863 (2007).

63. Yan, N., O'Day, E., Wheeler, L. A., *et al.* HIV DNA is heavily uracilated, which protects it from autointegration. *Proc Natl Acad Sci U S A* **108**, 9244–9249 (2011).

64. Bull, C. F., Mayrhofer, G., O'Callaghan, N. J., *et al.* Folate deficiency induces dysfunctional long and short telomeres; both states are associated with hypomethylation and DNA damage in human WIL2-NS cells. *Cancer Prev Res (Phila)* **7**, 128–138 (2014).

65. Montpetit, A. J., Alhareeri, A. A., Montpetit, M., *et al.* Telomere Length: A Review of Methods for Measurement. *Nurs Res* **63**, 289–299 (2014).

66. Cawthon, R. M. Telomere measurement by quantitative PCR. *Nucleic Acids Res* **30**, e47 (2002).

67. Cawthon, R. M. Telomere length measurement by a novel monochrome multiplex quantitative PCR method. *Nucleic Acids Res* **37**, e21 (2009).

68. O'Callaghan, N. J., Fenech, M. A quantitative PCR method for measuring absolute telomere length. *Biol Proced Online* **13**, 3 (2011).

69. Jain, M., Fiddes, I. T., Miga, K. H., *et al.* Improved data analysis for the MinION nanopore sequencer. *Nat Methods* **12**, 351–356 (2015).

70. Fleming, A. M., Ding, Y., Burrows, C. J. Sequencing DNA for the Oxidatively Modified Base 8-Oxo-7,8-Dihydroguanine. *Methods Enzymol* **591**, 187–210 (2017).

71. Goodwin, S., McPherson, J. D., McCombie, W. R. Coming of age: ten years of next-generation sequencing technologies. *Nat Rev Genet* **17**, 333–351 (2016).

72. Feng, Y., Cai, S., Xiong, G., Zhang, G., *et al.* Sensitive Detection of DNA Lesions by Bulge-Enhanced Highly Specific Coamplification at Lower Denaturation Temperature Polymerase Chain Reaction. *Anal Chem* **89**, 8084–8091 (2017).

73. Alsoe, L., Sarno, A., Carracedo, S., *et al.* Uracil Accumulation and Mutagenesis Dominated by Cytosine Deamination in CpG Dinucleotides in Mice Lacking UNG and SMUG1. *Sci Rep* **7**, 7199 (2017).

74. Nabel, C. S., Jia, H., Ye, Y., *et al.* AID/APOBEC deaminases disfavor modified cytosines implicated in DNA demethylation. *Nat Chem Biol* **8**, 751–758 (2012).

75. Ramsahoye, B. H., Burnett, A. K., Taylor, C. Nucleic acid composition of bone marrow mononuclear cells in cobalamin deficiency. *Blood* **87**, 2065–2070 (1996).

76. Sang, P. B., Srinath, T., Patil, A. G., *et al.* A unique uracil-DNA binding protein of the uracil DNA glycosylase superfamily. *Nucleic Acids Res* **43**, 8452–8463 (2015).

77. Andersen, S., Heine, T., Sneve, R., *et al.* Incorporation of dUMP into DNA is a major source of spontaneous DNA damage, while excision of uracil is not required for cytotoxicity of fluoropyrimidines in mouse embryonic fibroblasts. *Carcinogenesis* **26**, 547–555 (2005).

Chapter 9

Genomic Uracil — Valuable Tool in Molecular Biology but Inherent Problem in Sequencing of Ancient DNA

Geir Slupphaug* and Hans E. Krokan

Uracil holds a rather unique position among the non-canonical DNA bases in that it can be formed both enzymatically and spontaneously in DNA from cytosine. It can also be introduced via dUTP by DNA polymerases. Furthermore, it can be excised by DNA glycosylases to initiate DNA strand cleavage and nucleotide replacement. These characteristics provide opportunities to exploit uracil as a tool within molecular biology and medicine, but also represent inherent problems regarding DNA analysis, especially when sequencing ancient DNA. The following chapter will describe some analytical methods in which DNA-uracil constitutes an important intermediate. Finally, inherent problems associated with spontaneous cytosine deamination will be discussed in the context of analysis of stored archive samples and ancient DNA.

9.1. Use of Uracil and UDG to Control Carry-over Contamination in PCR

When PCR was implemented as a novel tool in molecular biology laboratories in the late 1980s, it soon became apparent that the spread

* *geir.slupphaug@ntnu.no*

of even trace amounts of amplified DNA fragments or primers in the laboratory environment could act as templates for the polymerase in new PCR reactions. Such carry-over contamination could lead to unintended amplification and false positive results, e.g. in clinical diagnostics.[1,2] To avoid this, many laboratories implemented safe-guarding protocols including separate rooms for PCR setup, UV germicidal lamps in biosafety hoods, separate sets of supplies as well as pipettes etc. In addition, positive and negative controls were implemented. In 1990, researchers at Life Technologies Inc. published two related uracil-PCR (UPCR) protocols aiming at eliminating the carry-over problem.[3] In the first protocol, dUTP replaced dTTP as building block in all PCR reactions. The products would then contain numerous dUMPs except in the primer regions. Addition of UDG during the assembly of all subsequent PCR reactions, would thus render carry-over DNA from previous reactions unamplifiable. Moreover, the UDG would be inactivated in the denaturation steps during thermal cycling so that new, dUMP-containing fragments would not be affected (Fig. 9.1). In the second protocol, only the PCR primers contained dUMP. The amplified DNA would thus contain dUMP in their 5′-regions only. Addition of UDG would nevertheless compromise amplification of such DNA. After the first denaturation step and inactivation of UDG, new dUMP-containing primers were then added to allow amplification of the desired target. The authors remarked that the presence of uracil in the PCR products should not affect most downstream usage, such as sequencing and hybridization, but that cloning of the products in bacteria would require use of a UDG-deficient host.

Most early UPCR protocols employed *E. coli* UDG (Ung) to excise the uracil. This UDG is relatively heat resistant, and thus residual UDG activity left after the first denaturation step often compromised the uracil-containing product. This was especially problematic for quantitative PCR (qPCR) and even more so for many reverse transcriptase PCR (RT-PCR) protocols, in which reverse transcription is performed under conditions similar to those utilized for UDG treatment.[4] Several attempts were thus made to generate more heat-labile UDGs,[5] which were employed in qPCR with variable results.[6,7]

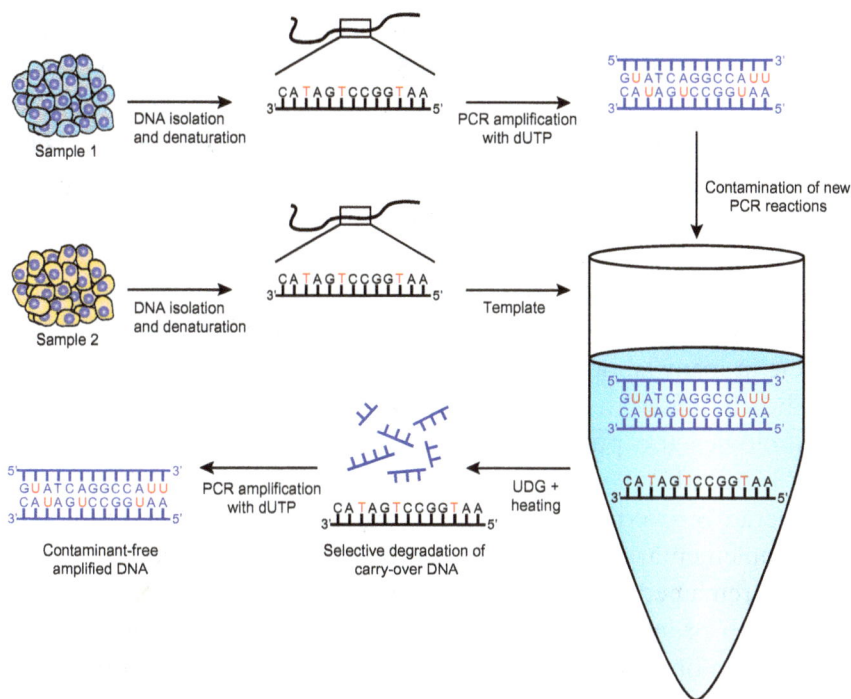

Fig. 9.1. Control of carry-over contamination in PCR by replacing dUTP for dTTP in all PCR reactions. Any amplified DNA entering into new PCR reactions can then be selectively degraded by addition of UDG to excise the uracils, followed by heating to cleave at abasic sites and inactivate the UDG.

One of these, cold-adapted UNG from Atlantic cod (*Gadus morhua*)[8] available from ArcticZymes®, has optimal activity at 37°C and is completely and irreversibly inactivated at 55°C, has proven to be very effective in many demanding RT-PCR settings.

9.2. Uracil-excision DNA Cloning, Sequencing and Engineering

Early site-specific mutagenesis protocols commonly suffered from low mutation frequencies, often arising from heteroduplex expression favouring the genotype of the original sequence. This was especially problematic when introducing silent mutations or mutations resulting in non-selectable phenotypes. In 1985, Thomas A. Kunkel published a method to specifically select against the original genotype, by producing the template DNA in an *E. coli dut⁻ ung⁻* strain.[9] These uracil-containing templates were then used to produce the mutated, uracil-free strand, and the ligated reaction mixture was treated with purified *E. coli* Ung and alkali to specifically degrade the template strand. This resulted in tenfold increased mutation frequencies (>80%) compared to other commonly used mutagenesis protocols. This method soon became routine in many laboratories and is still widely used.

Shortly after the publication of UPCR, researchers at Life Technologies also published the first method for uracil-based cloning.[10] Here, dUMP was incorporated into the 5′-end of PCR primers to generate 3′-overhangs after UDG treatment that could be annealed to complementary vector ends treated in the same fashion. The method remained largely unused for many years, due to its incompatibility with proof-reading DNA polymerases. Moreover, since UDG treatment alone did not mediate strand cleavage, the recombinant products were left with protruding single-stranded flaps that had to be processed *in vivo* by the host DNA repair machinery. At about the same time, several groups also published methods to improve sequencing of PCR products, by introducing one[11] or several[12] dUMP residues within the PCR primers and subsequent UDG treatment to produce single-strand DNA overhangs. In these reports, cleavage at the abasic sites formed after UDG treatment was promoted by heat or alkali treatment. In 2003, an improved method named USER™ (uracil-specific excision reagent) was commercialized by New England Biolabs, solving the problem of protruding single-strand flaps. This highly efficient ligase-free cloning strategy depended on the ability of 8 nt 3′-overhangs at both the PCR fragment and the vector to stably hybridize. The overhangs were generated by UDG-mediated uracil excision as in the original method, but were further processed by inclusion of the DNA glycosylase/lyase Endo VIII to mediate strand cleavage and dissociation of the region upstream of the cleavage site.[13] By varying the design of the PCR primers, the protocol could easily be adapted to perform simultaneous DNA manipulations such as directional cloning, site-specific mutagenesis, sequence insertion or deletion and sequence assembly. The incompatibility with proofreading DNA polymerases was partially resolved by the development of modified high-fidelity polymerases, like PfuTurboCX and PfuX7.[14–17] In contrast to unmodified Pfu that stalls at DNA template uracils, the modified enzymes were able to read through the uracils and thus to utilize uracil-containing primers to amplify DNA fragments with high fidelity. Improved vector systems were also introduced that increased robustness and resulted in improved efficiency, and thus rendering the method suitable for high-throughput cloning purposes. The USER method has also proven useful in high-throughput synthetic biology as a highly versatile and sequence-independent DNA assembly tool that facilitates simple site-directed mutagenesis and complex gene assemblies in simple one-tube protocols.[18–20] It has also been adapted to allow recombinant expression of multiprotein complexes in an insect cell baculoviral system and to produce large protein complexes containing more than 20 subunits.[21]

Uracil excision has been employed in an ingenious variant of the proximity ligation assay (PLA) to detect protein-DNA interactions (PDIs) in single cells with high specificity and sensitivity.[22] PLA detection of PDIs generally involves formation of a circular DNA molecule that can be amplified by rolling-circle amplification (RCA) to generate multiple targets that can be detected with fluorescence-labelled oligonucleotides. The circular DNA can be generated by different methods, each dependent on distinct DNA molecules hybridized to the DNA target and another DNA molecule bound to a secondary antibody recognizing the primary antibody bound to the target protein. Here, the group of Ola Söderberg at the Uppsala University obtained unsurpassed selectivity of the assay by hybridizing a modified padlock probe, containing two uracil-containing hairpins, to the target DNA. These hairpins prevented premature binding of the padlock probe to the PLA probe, and allowed the padlock probe to be ligated at elevated temperature by a thermostable Amp-ligase, thus decreasing false positive binding. Subsequent treatment of the uracil-containing padlock probe with UNG/EndoIV then removed the hairpins, allowing binding to the PLA-probe and formation of a circular DNA that could be amplified by RCA. An additional advantage of their strategy was that by omitting UNG, the total number of DNA targets in the cell could be identified, allowing calculation of binding stoichiometry of the protein of interest.[22]

9.3. Programmable DNA Editing by Targeted Cytosine Deamination

The CRISPR/Cas9 system is as a precise tool to introduce DNA DSBs as a first step in gene modification in mammalian cells.[23,24] A CRISPR/Cas9 protein-RNA complex localizes to a specific DNA sequence via base pairing with a guide RNA, and introduces a double-strand DNA break (DSB) at the locus specified by the guide RNA. As a response to the DSB, cells mobilize different DSB repair mechanisms, of which non-homologous end joining (NHEJ) mostly results in random insertions or deletions (indels) that frequently result in loss of functional protein expression. For most human genetic diseases, gene inactivation is not the goal, but rather the correction of inherited point mutations. Researchers have thus aimed at suppressing NHEJ, and to instead replace a mutated gene segment flanking the CRISPR/Cas9-induced DSB by homologous recombination with a wild-type sequence. However, under clinically relevant conditions such strategies to correct point mutations have been disappointingly inefficient.[25,26] To address this, Harvard researchers recently fused a mutant version of Cas9 (dCas9) that is deficient in double-strand DNA cleavage, to APOBEC1, and found that the fusion protein was capable of effectively introducing site-directed genomic C to U deamination *in vitro*, thereby effecting C→T (or G→A) substitutions.[27] To avoid repair of the generated uracil back to cytosine and thus extend the method to *in vivo* conditions, the researchers further fused Ugi to the dCas9-APOBEC1 to inhibit UNG-mediated excision of the deaminated cytosine (Fig. 9.2). This second-generation editor resulted in gene conversion efficiencies of up to 20% of total DNA sequenced. Moreover, by avoiding the DSB formation, the frequency of indels was found to be ≤0.1%. To increase base editing

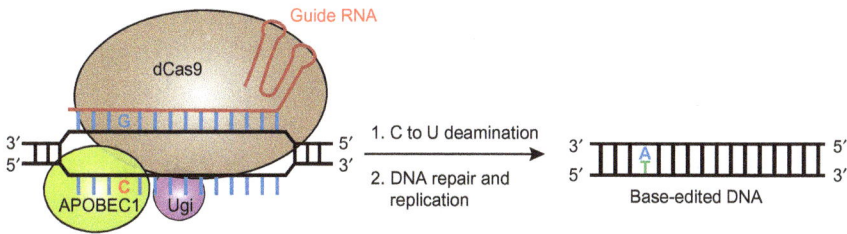

Fig. 9.2. A base pair editing fusion protein composed of modified Cas9, the cytidine deaminase APOBEC1 and the UNG-inhibitor Ugi converts C to U in a G:C base pair targeted by the guide RNA. Cellular DNA repair and replication convert the resulting G:U mismatch to an A:T (or A:U, not shown) base pair. (Modified from Ref. 27.)

further, they then mutated dCas9 further to nick the non-edited DNA strand opposite the U. This nick activates mismatch repair MMR to resolve the U:G pair. This additional modification resulted in 37% of the total DNA sequences containing the targeted C→T conversion and gave about 1% indels.[27] Shortly thereafter, several other groups published variants of the targeted deaminase approach, including the use of AID as deaminase, which increased the spectrum of mutations obtained.[28–31] The results strongly indicate that this technique may become a powerful tool to revert point mutations associated with human disease, including inherited disease, Alzheimer disease and cancer.

9.4. Ancient DNA — Rich Possibilities and Technical Limitations

The power of high throughput sequencing has changed genomics to a degree that could hardly be anticipated only two decades ago. Cancer genomics has now given extensive insights into the complex genetic changes that occur in cancer. High throughput sequencing has also made possible analyses of genetic diversity in extant human populations and has complemented archeology and paleoanthropology in historical questions about occurrence, disappearance and migration of populations. Even more challenging; it has given information about admixture between modern man and our closest known relatives, the Neanderthals and Denisovans.[32–34] Generally, the yields of ancient DNA are low, requiring PCR amplification prior to sequencing. This generates two types of problems; first, polymerase blocking lesions in DNA may strongly reduce PCR amplification; secondly, miscoding lesions result in false sequence information.[35]

After an organism is dead, the DNA can no longer be repaired. It is then being degraded by nucleases that are now not confined to cellular compartments, as well as by microorganisms that invade and degrade the organism. Thus, degradation of the DNA and contamination with DNA from other sources make analyses of ancient DNA difficult. Frequently, more than 99% of the DNA isolated from a sample of interest may be contaminating DNA. Furthermore, DNA is also fragmented and damaged due to its inherent chemical instability.[36] The quantitatively most frequent damage is hydrolytic depurination, resulting in loss of guanine and adenine from DNA. The resulting abasic sites block replicative polymerases. Numerous oxidative lesions in DNA bases, as well as in the backbone are known. Some of these block polymerase-catalyzed DNA synthesis, whereas others are miscoding. Various types of hydantoin lesions are quantitatively dominant in ancient DNA and appear to prevent PCR amplification, indicating that they are blocking lesions, or that other lesions correlate with the presence of blocking lesions.[37] Furthermore, analyses of ancient DNA have indicated that there are several blocking lesions that have not been characterized.[35]

Spontaneous hydrolytic deamination of cytosine to uracil is also frequent and will result in apparent C to T substitutions in DNA sequencing. Uracil in a U:G mismatch is miscoding and not blocking, a dangerous combination in PCR amplification. It may cause false results in DNA sequences in both stored archive material, as well as ancient DNA. Furthermore, cytosine deamination has been documented as an important source of apparent genetic substitutions in ancient DNA because pretreatment of such DNA samples with uracil-DNA glycosylase reduces the number of observed substitutions.[35,38–40] The increase in genomic uracil in DNA has also been turned into an advantage in a method for selective enrichment of uracil-containing

Neanderthal DNA. This method allowed ~10-fold increase in sequence information and some four- to five-fold yield in endogenous DNA sequences relative to those of microbial contaminants in some samples.[41]

The spontaneous decay of DNA due to hydrolytic depurination and depyrimidination, strand breaks, as well as cytosine deamination, follows the rules of chemistry and their rates have been determined.[36,42–45] It can therefore be calculated that the cumulative effects of such damage preclude the possibility of determining DNA sequences in animals living tens of millions years ago, such as dinosaurs. Rather, 100,000 years would appear to be enough to degrade essentially all DNA at physiological salt concentration, neutral pH and 15°C.[36,40,46] Successful and reproducible recovery of ancient DNA has been carried out from bone that shelters DNA from microorganisms, in a dry cold environment. Even then, yields are low and require great precautions and several types of controls to avoid false results. Claims of determining DNA sequences from DNA several million years old have so far not been reproduced.[40] However, measurement of the half-life of DNA in bone has indicated that at temperatures well below freezing, DNA of a size of 30 nt might survive possibly as much as one million years.[47] It is unlikely that conditions like that may frequently encountered in field studies, but targeted searches for ancient DNA protected by bone and low temperatures remain a possibility. In conclusion, DNA in dead organisms is subject to degradation by microorganisms, released nucleases, as well as spontaneous chemical decay. Cytosine deamination represents a particular problem since it causes errors in DNA sequencing. This problem may be eliminated, or at least reduced, by treating isolated DNA with uracil-DNA glycosylase. Therefore, high quality sequencing of ancient DNA from man and other organisms is of very high interest in interdisciplinary studies on the history of living organisms.

References

1. Gibbs, R., Chamberlain, J. The polymerase chain reaction: a meeting report. *Genes Develop* **3**, 1095–1098 (1989).

2. Kwok, S., Higuchi, R. Avoiding false positives with PCR. *Nature* **339**, 237–238 (1989).

3. Longo, M. C., Berninger, M. S., Hartley, J. L. Use of uracil DNA glycosylase to control carry-over contamination in polymerase chain reactions. *Gene* **93**, 125–128 (1990).

4. Thornton, C. G., Hartley, J. L., Rashtchian, A. Utilizing uracil DNA glycosylase to control carryover contamination in PCR: characterization of residual UDG activity following thermal cycling. *BioTechniques* **13**, 180–184 (1992).

5. Sobek, H., Schmidt, M., Frey, B., Kaluza, K. Heat-labile uracil-DNA glycosylase: purification and characterization. *FEBS Lett* **388**, 1–4 (1996).

6. Taggart, E. W., Carroll, K. C., Byington, C. L., *et al.* Use of heat labile UNG in an RT-PCR assay for enterovirus detection. *J Virol Methods* **105**, 57–65 (2002).

7. Pierce, K. E., Wangh, L. J. Effectiveness and limitations of uracil-DNA glycosylases in sensitive real-time PCR assays. *BioTechniques* **36**, 44–46, 48 (2004).

8. Lanes, O., Guddal, P. H., Gjellesvik, D. R., Willassen, N. P. Purification and characterization of a cold-adapted uracil-DNA glycosylase from Atlantic cod (Gadus morhua). *Comp Biochem Physiol B Biochem Mol Biol* **127**, 399–410 (2000).

9. Kunkel, T. A. Rapid and efficient site-specific mutagenesis without phenotypic selection. *Proc Natl Acad Sci U S A* **82**, 488–492 (1985).

10. Nisson, P. E., Rashtchian, A., Watkins, P. C. Rapid and efficient cloning of Alu-PCR products using uracil DNA glycosylase. *PCR Methods Appl* **1**, 120–123 (1991).

11. Day, P. J., Walker, M. R. Sequencing self-ligated PCR products using 3′ overhangs generated by specific cleavage of dUTP by uracil-DNA glycosylase. *Nucleic Acids Res* **19**, 6959 (1991).

12. Ball, J. K., Desselberger, U. The use of uracil-N-glycosylase in the preparation of PCR products for direct sequencing. *Nucleic Acids Res* **20**, 3255 (1992).

13. Bitinaite, J., Rubino, M., Varma, K. H., *et al.* USER friendly DNA engineering and cloning method by uracil excision. *Nucleic Acids Res* **35**, 1992–2002 (2007).

14. Bitinaite, J., Nichols, N. M. DNA cloning and engineering by uracil excision. *Curr Protoc Mol Biol* **Chapter 3**, Unit 3 21 (2009).

15. Geu-Flores, F., Nour-Eldin, H. H., Nielsen, M. T., Halkier, B. A. USER fusion: a rapid and efficient method for simultaneous fusion and cloning of multiple PCR products. *Nucleic Acids Res* **35**, e55 (2007).

16. Norholm, M. H. A mutant Pfu DNA polymerase designed for advanced uracil-excision DNA engineering. *BMC Biotechnol* **10**, 21 (2010).

17. Nour-Eldin, H. H., Geu-Flores, F., Halkier, B. A. USER cloning and USER fusion: the ideal cloning techniques for small and big laboratories. *Methods Mol Biol* **643**, 185–200 (2010).

18. Nielsen, M. T., Ranberg, J. A., Christensen, U., *et al.* Microbial synthesis of the forskolin precursor manoyl oxide in an enantiomerically pure form. *Appl Environ Microbiol* **80**, 7258–7265 (2014).

19. Nielsen, M. T., Madsen, K. M., Seppala, S., *et al.* Assembly of highly standardized gene fragments for high-level production of porphyrins in E. coli. *ACS Synth Biol* **4**, 274–282 (2015).

20. Cavaleiro, A. M., Kim, S. H., Seppala, S., *et al.* Accurate DNA assembly and genome engineering with optimized uracil excision cloning. *ACS Synth Biol* **4**, 1042–1046 (2015).

21. Zhang, Z., Yang, J., Barford, D. Recombinant expression and reconstitution of multiprotein complexes by the USER cloning method in the insect cell-baculovirus expression system. *Methods* **95**, 13–25 (2016).

22. Weibrecht, I., Gavrilovic, M., Lindbom, L., *et al.* Visualising individual sequence-specific protein-DNA interactions in situ. *N Biotechnol* **29**, 589–598 (2012).

23. Cho, S. W., Kim, S., Kim, J. M., Kim, J. S. Targeted genome engineering in human cells with the Cas9 RNA-guided endonuclease. *Nat Biotechnol* **31**, 230–232 (2013).

24. Mali, P., Yang, L., Esvelt, K. M., *et al.* RNA-guided human genome engineering via Cas9. *Science* **339**, 823–826 (2013).

25. Cong, L., Ran, F. A., Cox, D., *et al.* Multiplex genome engineering using CRISPR/Cas systems. *Science* **339**, 819–823 (2013).

26. Ran, F. A., Hsu, P. D., Wright, J., *et al.* Genome engineering using the CRISPR-Cas9 system. *Nat Protoc* **8**, 2281–2308 (2013).

27. Komor, A. C., Kim, Y. B., Packer, M. S., *et al.* Programmable editing of a target base in genomic DNA without double-stranded DNA cleavage. *Nature* **533**, 420–424 (2016).

28. Nishida, K., Arazoe, T., Yachie, N., *et al.* Targeted nucleotide editing using hybrid prokaryotic and vertebrate adaptive immune systems. *Science* **353** (2016).

29. Ma, Y., Zhang, J., Yin, W., *et al.* Targeted AID-mediated mutagenesis (TAM) enables efficient genomic diversification in mammalian cells. *Nat Methods* **13**, 1029–1035 (2016).

30. Hess, G. T., Fresard, L., Han, K., *et al.* Directed evolution using dCas9-targeted somatic hypermutation in mammalian cells. *Nat Methods* **13**, 1036–1042 (2016).

31. Yang, L., Briggs, A. W., Chew, W. L., *et al.* Engineering and optimising deaminase fusions for genome editing. *Nat Commun* 7, 13330 (2016).

32. Fu, Q., Hajdinjak, M., Moldovan, O. T., *et al.* An early modern human from Romania with a recent Neanderthal ancestor. *Nature* **524**, 216–219 (2015).

33. Prufer, K., Racimo, F., Patterson, N., *et al.* The complete genome sequence of a Neanderthal from the Altai Mountains. *Nature* **505**, 43–49 (2014).

34. Sankararaman, S., Mallick, S., Dannemann, M., *et al.* The genomic landscape of Neanderthal ancestry in present-day humans. *Nature* **507**, 354–357 (2014).

35. Dabney, J., Meyer, M., Paabo, S. Ancient DNA damage. *Cold Spring Harb Perspect Biol* **5** (2013).

36. Lindahl, T. Instability and decay of the primary structure of DNA. *Nature* **362**, 709–715 (1993).

37. Hoss, M., Jaruga, P., Zastawny, T. H., *et al.* DNA damage and DNA sequence retrieval from ancient tissues. *Nucleic Acids Res* **24**, 1304–1307 (1996).

38. Paabo, S. Ancient DNA: extraction, characterization, molecular cloning, and enzymatic amplification. *Proc Natl Acad Sci U S A* **86**, 1939–1943 (1989).

39. Hofreiter, M., Jaenicke, V., Serre, D., *et al.* DNA sequences from multiple amplifications reveal artifacts induced by cytosine deamination in ancient DNA. *Nucleic Acids Res* **29**, 4793–4799 (2001).

40. Hofreiter, M., Serre, D., Poinar, H. N., *et al.* Ancient DNA. *Nat Rev Genet* **2**, 353–359 (2001).

41. Gansauge, M. T., Meyer, M. Selective enrichment of damaged DNA molecules for ancient genome sequencing. *Genome Res* **24**, 1543–1549 (2014).

42. Lindahl, T., Andersson, A. Rate of chain breakage at apurinic sites in double-stranded deoxyribonucleic acid. *Biochemistry* **11**, 3618–3623 (1972).

43. Lindahl, T., Karlstrom, O. Heat-induced depyrimidination of deoxyribonucleic acid in neutral solution. *Biochemistry* **12**, 5151–5154 (1973).

44. Lindahl, T., Nyberg, B. Rate of depurination of native deoxyribonucleic acid. *Biochemistry* **11**, 3610–3618 (1972).

45. Lindahl, T., Nyberg, B. Heat-induced deamination of cytosine residues in deoxyribonucleic acid. *Biochemistry* **13**, 3405–3410 (1974).

46. Lindahl, T. Recovery of antediluvian DNA. *Nature* **365**, 700 (1993).

47. Allentoft, M. E., Collins, M., Harker, D., *et al.* The half-life of DNA in bone: measuring decay kinetics in 158 dated fossils. *Proc Biol Sci* **279**, 4724–4733 (2012).

Index